Boredom, *Shanzhai*, and Digitisation in the Time of Creative China

Boredom, *Shanzhai*, and Digitisation in the Time of Creative China

Edited by
Jeroen de Kloet, Chow Yiu Fai, and Lena Scheen

Routledge
Taylor & Francis Group
LONDON AND NEW YORK

The publication of this book is supported by a consolidator grant from the European Research Council (ERC-2013-CoG 616882-ChinaCreative).

First published in 2019 by Amsterdam University Press Ltd.

Published 2025 by Routledge
4 Park Square, Milton Park, Abingdon, Oxon OX14 4RN
605 Third Avenue, New York, NY 10158

Routledge is an imprint of the Taylor & Francis Group, an informa business

ISBN: 9789462984745 (hbk)
ISBN: 9781041176350 (pbk)
ISBN: 9781003691839 (ebk)
NUR 630

Cover illustration: Jeroen de Kloet

Cover design: Coördesign, Leiden

DOI: 10.5117/9789462984745

For Product Safety Concerns and Information please contact our EU representative:
GPSR@taylorandfrancis.com
Taylor & Francis Verlag GmbH, Kaufingerstraße 24, 80331 München, Germany

Table of Contents

Notes from the Field

In Dialogue

Essay

Interlocuting

Notes from the Field

In Dialogue

List of Illustrations

Acknowledgments

This book is originated from a workshop titled *You Must Create? Boredom,* Shanzhai *and Digitisation in the Time of 'Creative China'*. Held in Hong Kong from 7-9 April 2016, the workshop was jointly organised by the Amsterdam Centre for Globalization Studies, University of Amsterdam and the Department of Humanities and Creativity Writing, Hong Kong Baptist University, as part of the European Research Council funded project 'From Made in China to Created in China – A Comparative Study of Creative Practice and Production in Contemporary China' (ChinaCreative, grant number 616882). We want to thank the organisers, contributors and funders of the workshop. We also want to thank the Amsterdam University Press – especially Saskia Gieling, Mike Sanders, Jaap Wagenaar and Anna Yeadell-Moore – for their professional support and collegial encouragement. Connie Cheng was totally indispensable in coordinating the workshop and subsequently the book project. Gordon So was always reliable when we needed last-minute house style update. And the three of us want to thank our respective affiliations for affording us to complete this book. Finally, yes, we want to thank each other – it was so much fun!

Jeroen de Kloet, University of Amsterdam; Communication University of China
Chow Yiu Fai, Hong Kong Baptist University
Lena Scheen, New York University Shanghai

May 2019

I Introduction

We Must Create?

Jeroen de Kloet, Chow Yiu Fai, and Lena Scheen

Abstract

Premised on the *imperative of creativity*, this introduction explicates discourses saturated with notions of individual talent marked by inspiration, newness, and innovation. It offers context to the book's aim to critically re-enter the idea of creativity, by aligning it to three concepts, considered by the authors to be emblematic for the Creative China of our time: boredom, *shanzhai* (a vernacular Chinese term connoting copying and appropriation), and digitisation. All three hover around ideas of innovation, newness, and their constitutive flip sides of copying and repetition. After a discussion of these concepts, foregrounding their concerns and the question of what creativity may enable as well as disable, the chapter introduces the organising logic of the book and the ensuing chapters.

Keywords: creativity, Creative China, boredom, *shanzhai*, digitisation

> The future no longer seems to promise anything fundamentally new; instead, we imagine endless variations on what already exists.[1]

The imperative of creativity

The collaborative writing process of this introduction was like a rather strenuous, winding journey full of detours and distractions. While waiting inspiration, we would often go online, check our Facebook, or use WeChat to chat with one another and send some pictures. This writing process,

1 Boris Groys, *On the New*, trans. G.M. Goshgarian (Brooklyn: Verso, 2014 [1992]), 1.

Kloet, Jeroen de, Chow Yiu Fai, and Lena Scheen (eds), *Boredom,* Shanzhai*, and Digitisation in the Time of Creative China*. Taylor & Francis Group, 2019
DOI: 10.5117/9789462984745_INTRO

likely quite familiar to the reader, is often a fight with or against boredom, as searching for the right angle for the text, the right opening, the best structure, goes slowly and is, above all, a messy affair. Waiting for inspiration too often turns out to be like *Waiting for Godot* – it is destined never to arrive, really. Instead, it seems wiser to use an app that carries the apparently unintentionally ironic name of *Freedom*, which shuts down your internet connection, and then to just start writing, trying to ignore that feeling that haunts so much of our writing: it has to be something new! And yet, so much has already been written! We cannot and should not copy things we have already written, and, on top of that quite impossible demand, how to add something to the massive body of literature that has already been published?

These struggles, we believe, are not unique to the three of us, as they haunt both academic and creative work. What we like to call the *imperative of creativity* in this introduction, drags us into discourses that are saturated with notions of individual talent that is marked by inspiration, newness, and innovation. These celebrated notions often operate as disciplinary straightjackets; they confine rather than liberate, to the extent that they may even paralyse our writing. How to liberate creativity from these straightjackets? This book is an attempt to critically re-enter the idea of creativity, hopefully to steer away from its disciplinary implications, and we do so by aligning it to three different and interconnected concepts: boredom, *shanzhai* (a vernacular Chinese term connoting copying and appropriation), and digitisation. As our short reflection upon our writing process already alludes to, these three notions are intimately entangled with creativity: the struggle of writing is often also a struggle with and against boredom, one that is dovetailed by a fear of copying, of writing nothing new. As Pang Laikwan writes, 'the force of the new has always driven modernity – in the social discourse of modernisation and technological development, and in the aesthetic discourse of modernist arts'.[2] Today, our digital culture offers both a welcome escape route, but also makes the process of writing even more hazardous, as texts and information have amplified – we may well label the internet a loquacious device – to the point that we hardly know where to start researching and archiving, making the demand for the new seem all the more out of reach.

Boris Groys continues the line with which we open this introduction as follows, 'For many people, it is depressing to imagine the future as an endless reproduction of the past and present. For others, a new age in social and artistic practice is dawning, one liberated from the dictates of the new and from diverse

2 Lai-kwan Pang, '"China Who Makes and Fakes": A Semiotics of the Counterfeit', *Theory, Culture & Society* 25.6 (2008): 117-40.

future-oriented utopian and totalitarian ideologies.' He wrote these words in 1992, at the heyday of poststructuralist and postmodern thinking, a time in which any claim towards either the new, or the utopian future would meet with suspicion and disbelief. While we are less mesmerised by the postmodern today, we do want to commit ourselves in this book to its questioning of the new, or maybe better, we wonder: can we rearticulate the new in such a way that it does not devalue ideas of copying, of repetition, and of mimicry?

The imperative of creativity is not restricted to the domain of cultural production. It permeates all professional and private realms: a manager, a scholar, a teacher, a parent, or a child all ought to be creative today. Self-help handbooks are written on how to foster your creativity; take, for example, Rod Judkins' *The Art of Creative Thinking*, a book that is 'passionately about taking the spirit of creativity that exists in the art world out into the wider world'.[3] Throughout the book, quotes are given from the usual, and of course Western, creative suspects, ranging from Salvador Dali to Daniel Radcliffe and from George Eliot to Arthur Koestler. The latter is quoted saying that 'creativity is the defeat of habit by originality',[4] a statement once again validating the demand for the new and the original. It is therefore not surprising that all around the world, city governments embrace the idea of the creative city and emerging economies jump on the same bandwagon in order to move from a place of manufacturing to a place of creation. This slowly pushes us towards the paradoxical situation in which all cities are creative – a situation in which creativity is bound to evaporate as a meaningful category. In creativities' slipstream, other words like originality, innovation, smart, and sustainability often follow suit. China is part of this global trend. In this book, we want to steer away from uncritically celebrating the emergence of what Angela McRobbie aptly calls the creativity dispositif.[5] At the same time, we feel uncomfortable with its inversion, the critique, often inspired by neo-Marxism, which reads creativity solely as a driving force for a global neo-liberalism that produces precarious jobs and propels further inequalities between classes as well as between places.[6]

We straddle these two positions, searching both for what creativity may enable as well as disable. We do so by way of probing, as creativity 'has

3 Rod Judkins, *The Art of Creative Thinking: 89 Ways to See Things Differently* (New York: Tarcher Perigee, 2016), 2.

4 Ibid., 234.

5 Angela McRobbie, *Be Creative: Making a Living in the New Culture Industries* (Cambridge, UK; Malden, MA: Polity, 2016).

6 For this critique, see, among others, McRobbie, *Be Creative*, and Isabell Lorey, *State of Insecurity: Government of the Precarious*, trans. Aileen Derieg (London; New York: Verso, 2015).

a strong tendency to resist definition',[7] in particular the proliferation of different creativities in China – to see how they resonate and dissonate with experiences elsewhere. Seeing this as a journey with an unclear destiny, afraid as we are of overcoded and univocal arguments that are often driven by fixed binaries and strong and rigid ideological positions, we have tried to include different voices: those of PhD students who were in the midst of their fieldwork when invited to send in their tentative findings and thinking; those of scholars already engaged for a longer time with the issue of creativity and of Chinese creativity; and those of artists and curators, practitioners working in the creative fields. We must probably thank and blame the imperative of creativity for guiding us to edit the book this way.

In the essay following this introduction, Eitan Wilf traces a genealogy of the idea of creativity from the Greeks' divine inspiration, via the romantic idea of individual genius to its current commoditised and copyrighted version. His chapter, together with this introduction, constitutes the general framing of this book. The subsequent three parts are organised according to three interrelated phenomena that we consider emblematic for the Creative China of our time: boredom, *shanzhai*, and digitisation.

First, as already alluded to in our opening to this introduction, our lives are continuously haunted by the spectre of boredom, of monotony, of doing the same thing time and again. The daily routine of a nine-to-five (or, more likely, later than five) job, the writing of yet another academic article, household chores, the perfunctory performance of bouquets and chocolates at anniversaries – all seem to be entrenched in quotidian fixities that make us feel helpless and uncreative. Creativity may hold the promise here of an escape from boredom, of doing things and living lives differently, and of thrusting open an aspirational window on the future. But boredom may also be an important constitutive condition for creativity.

Second, the notion of *shanzhai* allows us to go deeper into, and question, the already mentioned notions like new, innovation, unique, and talented. The emergence of *shanzhai* culture in China, in which cities, buildings, phones, people, books, and so forth, are being mimicked, is often read as an infringement of copyright law – the juridical underpinning of the reading of creativity as something unique, individual (or better, locatable), and innovative. But *shanzhai* practices attest to something different: the 'fake' iPhone has more rather than fewer functions, while new items are added to the 'real' Paul Smith winter collection. In general, it has helped to engender a vernacular culture of copying and pasting that is embraced by many Chinese

7 Pang, '"China Who Makes and Fakes"', 118.

youngsters. Or think of the mimetic practice in calligraphy, where an ideal is to write like your master, to become a perfect imitator. *Shanzhai* cultures may also help to revalidate the importance of craftsmanship, as the focus is more on making than on creating. Amidst this diversity, we are interested in the contingency and potentiality in *shanzhai* practices.

Third, digitisation is a force that is currently transforming the global cultural and creative landscape at rapid speed. As elsewhere in the world, China aspires to become a leading nation in new technology and has initiated its Internet Plus policy as a driver for economic growth. At the same time, the penetration of new technologies in everyday life seems to surpass that in the West: *Weixin*, in addition to its Whatsapp-ish functions, is fast morphing into a financial tool, while the new generation's current career dream is to develop an app. New technologies also pose a challenge to traditional forms of creative education and legitimisation: nowadays, one can pick up a digital recorder, or say an iPhone, to shoot a movie, regardless of whether or where one is trained. Concomitantly, it also generates possibilities of 'citizen journalism', particularly relevant to a heavily controlled media landscape like that in China.

These brief accounts, a prelude to the remains of this introduction, concerning the intersections between boredom, *shanzhai*, digitisation, and creativity as played out in contemporary China, await elaborations, examinations, and problematisations. These three aspects may strike the reader as rather haphazard, but we contend that they are not. All three hover around ideas of innovation, newness, and their constitutive flip sides of copying and repetition. Boredom alludes to the repetitive in everyday life, and thus is perceived to be at odds with innovation and renewal. *Shanzhai* culture refers to the culture of the fake that proliferated so abundantly in China. As this volume will show, the culture of the fake does not exclude creativity, but rather may give it an important twist. Finally, the realm of the digital allows for the endless copying and circulation of cultural practices, once again pushing the dialectics of repetition and creation further. Thinking about the three aspects of boredom, *shanzhai*, and digitisation together, we argue, will help us to reimagine the imperative of creativity, in particular its driving forces of the new, the talented, and the original.

Thus, we wonder in this volume, how do boredom, *shanzhai*, and digitisation impact on creative practices in China, and vice versa? What is actually going on at the interface of creative and everyday lives? How (far) do 'Chinese' creative practices, as understood from the three interrelated phenomena, facilitate different ways of theorising and understanding 'creativity'? Before embarking on different ways to answer these questions, let us take a closer look at the three guiding concepts of this volume.

Boredom

We were never being bored
'Cause we were never being boring
'Being Boring', Pet Shop Boys

These catchy lines, for the generation growing up with the electronic duo (including the authors of this chapter), have become some kind of mantra performing a paradoxical function. It celebrates the fun, the beauty, the youth of a party-going life and, at the same time, it laments its inevitable demise, when the parties end and the youthful grows into, to use another Pet Shop Boys hit, the suburbia of boring adulthood. 'Being Boring' was released in 1990. Five years earlier, educator and communication theorist Neil Postman published a book outlining a dystopic world ruled by the logic of pleasure, entertainment, and triviality. As enshrined in the alarming title – *Amusing Ourselves to Death: Public Discourse in the Age of Show Business* – this new world thrives on the anxiety of being bored and boring, and turns itself, at least according to Postman, into something worse than the dictatorial and pain-inflicting regime of *1984*. Indeed, citing in particular television and TV news in his time, Postman argues persuasively that it is not George Orwell's *1984* that we should fear, but Aldous Huxley's *Brave New World*. 'What Orwell feared were those who would ban books. What Huxley feared was that there would be no reason to ban a book, for there would be no one who wanted to read one'.[8]

The Pet Shop Boys' song and Postman's book differ in their central concern: youth ideals and entertainment culture, respectively. What they share is a sensitivity towards something happening in the 1980s and the 1990s they were experiencing. For one, it is the anxiety of boredom; for the other, it is the anxiety of that anxiety. Taken together, it is hardly surprising that boredom is often conceived as a modern affect, discourse, and invention; something, say, of our time. We become bored when the modern condition ushers us to attach increasing importance to leisure, to feel that we should be happy, to have largely abandoned religious preoccupations, to increasingly function individualistically, and to live in a standardisation of time and space.[9] In

8 Niel Postman, *Amusing Ourselves to Death: Public Discourse in the Age of Show Business* (London: Penguin Books, 2005 [1985]), xix.

9 Ben Anderson, 'Time-stilled Space-slowed: How Boredom Matters', *Geoforum* 35.6 (2004): 739-54.

the words of Elizabeth Goldstein, '[i]n a time when the drives to novelty and innovation, speed and progress that have always defined modernity have become the foundation of a process of continuously accelerating transformation, boredom haunts the Western world'.[10] Citing a number of philosophers (from Pascal to Adorno) and writers (from Goethe to Pessoa) on boredom, Lars Svendsen notes that all of them 'belong to the modern period'.[11] Some turn to language to argue for the modern emergence of boredom. They argue that the word 'boredom' has a relatively short history, emerging somewhere in the eighteenth century but only morphing to general usage decades later. 'To bore' is traced to the following century, in 1812 to be exact,[12] and 'boredom' to 1864.[13] Such genealogy is troubled by a quote from Søren Kierkegaard, who passed away in 1855, and is claimed to have expressed 'How dreadful boredom is – how dreadfully boring [...] I lie prostrate, inert; the only thing I see is emptiness, the only thing I live on is emptiness, the only thing I move in is emptiness'.[14] In response to such disqualifications, and making a leap in time towards the present, others have argued that boredom can also be considered a virtue. Brian O'Connor, for example, argues in *Idleness – A Philosophical Essay*, drawing in particular on German idealism and its aftermath, that 'while idleness involves a way of living that has ceded from social pressures, thereby reducing the scope of the influence of socialisation itself, it too may be understood as an expression of flourishing'.[15] He draws on, among others, Schiller's and Marcuse's claim on the importance of play in life. Idleness can be considered boredom's twin, and O'Connor's view helps us to redirect our gaze from sheer condemnation towards its more productive dimensions.

Taking boredom as a state of mind, literary scholar Patricia Meyer Spacks traces the contemporaneous rise of both the idea of boredom and the genre of the novel in the context of early modernity, particularly that of leisure. Another line of work that connects boredom to modernity is informed by practices of non- or pre-modern cultures. Anthropologist Yasmine

10 Elizabeth S. Goodstein, *Experience Without Qualities: Boredom and Modernity* (Stanford, CA: Stanford University Press, 2008), 1.

11 Lars Svendsen, *A Philosophy of Boredom*, trans. John Irons (London: Reaktion, 2005), 20.

12 Barbara Dalle Pezze and Carlo Salzani, 'Introduction: The Delicate Monster: Modernity and Boredom'. In *Essays on Boredom and Modernity*, eds. Barbara Dalle Pezze and Carlo Salzani (Amsterdam: Rodopi, 2008), 11.

13 Patricia Meyer Spacks, *Boredom: The Literary History of a State of Mind* (Chicago: University of Chicago Press, 1995), 32.

14 Søren Kierkegaard, *Either/Or: A Fragment of Life* (Copenhagen: University Bookshop Reitzel, 1843).

15 Brian O'Connor, *Idleness – A Philosophical Essay* (Princeton: Princeton University Press, 2018), 20.

Musharbash studies the life of aboriginal populations in Australia before colonisation.[16] Demonstrating how they were living in the moment, Musharbash believes that they did not experience boredom and their language had no word to such effect or affect. The colonisers, bringing with them clocks, other modernist gadgets, and conceptions of linear time – i.e. the past and memory – brought boredom to the aboriginal world. Such studies resonate with earlier thinkers reflecting on time and modernity. Henri Lefebvre, gesturing to Heidegger, considers life in pre-modern societies 'was organised in relation to the endless, undulating cycles of birth and death, remembrance and recapitulation that mark the natural world', generating 'constant newness within continuity'.[17] This marks a stark contrast with the modern time, or the time of the modern, organised by repetition, which 'is derived from the dictates of technology, work and production'.[18]

Are only modern human beings capable of being bored? Some wonder. Firstly, a host of studies in such disciplines as neuroscience, psychology, and psychiatry has sought to demonstrate the biological and thus universal nature of boredom. Drawing on these studies, philosopher Peter Toonley argues against the modern history of boredom. He is of the view that human beings have many ways to express such boredom without necessarily resorting to naming it directly. At the same time, he makes a useful distinction between two kinds of boredom, the quotidian and the existential, the latter of which seems to be what many scholars on boredom are reflecting on. If we are talking about the boredom we may experience on a daily basis, Toonley does not believe it is the privilege of the moderns. Instead, '[h]umans always have had the capacity for this emotion [of boredom]'; it is just that 'not all societies enable or require humans to experience boredom'.[19] Having established this premise, Toonley proposes to revisit boredom in the current society – he cites the credit crunch as a defining characteristic – and presents boredom as a useful emotion, particularly as 'an early warning signal that certain situations may be dangerous to our well-being'.[20]

Similar to Toonley, we want to focus on the quotidian kind of boredom that, we believe, shows the most immediate and interesting bearings with the general theme of this book: creativity. We place our inquiry in contemporary

16 Yasmine Musharbash, 'Boredom, Time, and Modernity: An Example from Aboriginal Australia', *American Anthropologist* 109.2 (2007): 307-17.

17 Michael E. Gardiner, 'Henri Lefebvre and the "Sociology of Boredom"', *Theory, Culture & Society* 29.2 (2012): 43.

18 Ibid., 44.

19 Peter Toohey, *Boredom: A Lively History* (New Haven, CT: Yale University Press, 2012), 156.

20 Ibid., 174.

China and seek to examine the intertwining of boredom with a society whose citizens are groomed with modern and cosmopolitan longings, aspirations, and needs;[21] whose youth negotiate surveillance and discipline with spaces opened up in the realms of entertainment and popular culture;[22] and whose state policy of soft power works in tandem with local and global capital to engender an ideology and proliferation of practices loosely described as creative.[23] In short, it is a society on the trajectory of modernity, but, so to speak, with Chinese characteristics, and our take on boredom here is how does this society of contemporary China enable or require its populations to experience boredom through creative practices?

Specifically, while we are not interested in taking boredom as a warning, or as something to celebrate and to lament, we follow many of the works cited earlier and take boredom not necessarily or exclusively as something negative; rather, we take it as a useful starting point to lodge two questions that are possibly one: how does creativity help mitigate boredom? And how does boredom incubate creativity? We are aware that these questions and assertions are not new and hasten to add that such a demand for the perpetually new actually runs counter to the spirit of this volume. Indeed, boredom and being idle is considered by a wide range of authors, ranging from philosophers to novelists, and from psychologists to writers of self-help books, to be constitutive of creativity. One example from the genre of self-help books comes from The School of Life, written by Eva Hoffman, with the telling title *How to be Bored*.[24] In this book, Hoffman suggests that modernity has gone berserk, ushering in a world of speed ridden with anxiety and hyperactivity and, like Neil Postman, it is the media, this time the digital media, that are one of the main culprits. The book offers an array of counter strategies to stimulate and foster boredom, ranging from introspection and idleness to creative play and keeping a diary.

In this volume, we are similar to and distinct from the aforementioned treatises on boredom. We are similar, for instance, to Michael Gardiner's re-entry into Lefebvre's 'sociology of boredom' when he points to Lefebvre's interest in the question of 'how the vague dissatisfactions we usually associate

21 Lisa Rofel, *Desiring China: Experiments in Neoliberalism, Sexuality, and Public Culture* (Durham, NC: Duke University Press, 2007).

22 Jeroen de Kloet and Anthony Y. H. Fung, *Youth Cultures in China* (Malden, MA: Polity, 2017).

23 Michael Keane, *Created in China: The Great New Leap Forward* (Abingdon: Routledge, 2007); *Creative Industries in China: Art, Design and Media* (Malden, MA: Polity, 2013); Francois Jullien, *The Silent Transformation*, trans. K. Fijalkowski and Michael Richardson (London: Seagull Books, 2011).

24 Eva Hoffman, *How to be Bored* (London: The School of Life, 2016).

with boredom can elide into other, more transformational instances'.[25] We are similar in striving for a better understanding of 'our human propensities for passionate engagement, play and the ludic [...], the pleasurable and the aesthetic'.[26] Nevertheless, our enquiries here are less oriented towards the existential, at least not in the utopian sense of the transformative potentials of boredom: 'a slow but profound modification of the everyday – of a new usage of the body, of time and space, of sociability; something that implies a social and political project'.[27] We agree with Ben Anderson that boredom matters but we do not align exactly with his thinking on 'movements-from boredom, such as joy, hope and despair' for Bloch's conception of the 'not-yet become'.[28] We are sceptical of uncritical celebrations of boredom as an antidote to a world of speed and anxiety, such as Hoffman's book. But her connection to hyperactivity and speed does gesture towards the case of China. In its dazzlingly rapid change, its compressed modernity, in its hyperactive and deeply digitised everyday life, China strikes us as a unique case. The difference with other places is less a matter of cultural particularity and more a matter of sheer intensity. When everyday life in the Chinese city resembles that of a pressure cooker, especially for the young generation,[29] what are the possible alliances and articulations between boredom and creativity? And between boredom, *shanzhai*, and digitisation? Our inquiries on creative practices in contemporary China may well shed light on these grander questions of humanity's and China's future; but our immediate remit is less philosophical than empirical, as we seek to discover how some populations on earth deal creatively with boredom in the now.

Reverting to the issue of language, the Chinese equivalent to the word 'boredom' may well be 悶: pictorially a heart 心 inside a door 門.[30] While some would read the imprisonment of the heart within a door as something akin to boredom, it would be equally persuasive to emphasize the heart beating to break out. What is boredom? It is the dynamic between the heart and the door, the politics of closing and opening that intrigues us. In other words, we do not deliberate on what boredom is, but what boredom *does*, and we do it in the field of creative practices in a modernising and globalising China.

25 Gardiner, 'Henri Lefebvre', 53.

26 Ibid.

27 Henri Lefebvre, 'Toward a Leftist Cultural Politics: Remarks Occassioned by the Centenary of Marx's Death', in *Marxism and the Interpretation of Culture*, eds. L. Grossberg and C. Nelson (Urbana, IL: University of Illinois Press, 1988), 86-7.

28 Anderson, 'Time-stilled Space-slowed', 740.

29 de Kloet and Fung, *Youth Cultures in China*.

30 Available at http://www.fantizi5.com/ziyuan/, accessed July 13, 2017.

In the section on boredom, Anneke Coppoolse connects boredom with the modern condition of commodification and consumer culture to tease out what she calls the 'aesthetics of overabundance' in the works of two artists based in Hong Kong. In her study of the traditional performance of *muyuge* in Dongguan, Wen Cuiyan examines what boredom does to this particular intangible cultural heritage amidst state-sanctioned nationalistic narratives and economic imperatives. Following these two essays, Esther Peeren zooms out from the Chinese context and offers an intervention in rethinking boredom and creativity through the Bored Panda website and adult colouring books. Invoking a photo of vertical gardening, which he took with his smartphone in Hong Kong, Christoph Lindner considers the digitisation of the image and the slowness of the imaged practice to question the culture of speed and connectivity, quite the antidote to boredom, on creative practices. As field notes from the public parks in Beijing, Laura Vermeeren observes and explores how elderly people practice an ephemeral form of art – water calligraphy – as a way to let the ennui of their retired life evaporate with words they write on the park grounds. Li Hao, on the other hand, outlines the emergence of a new television genre in China – the male Cinderella drama – as audiences are presumably bored by the conventional female prototype. Finally, artist Kingsley Ng and composer Kung Chi Shing engage in a dialogue on performativity in what the interviewer Lo Yin Shan calls the age of banality.

Shanzhai

> There is no such thing as too much *shanzhai* ~ only more *shanzhai*~
> The Chinese people are great
> Chinese people's creativity is limitless
> Long live the Chinese people
> Resolutely attack legitimate products, support pirate products
> *Comment by a Chinese netizen under a blogpost on* shanzhai *products*[31]

A life-sized virtual image of the Chinese student Fu Xin at the exhibition of the Chinese-American exchange programme 'Bringing the Chinese Dream to the U.S.' (made by his classmates after Fu was denied a visa to join the programme);

31 Available at https://www.chinasmack.com/funny-clever-chinese-shanzhai-brands, accessed July 13, 2017.

a smiling President Obama above the text 'Obama's BlackBerry. My BlockBerry Whirlwind 9500' on the advertisement of a knock-off BlackBerry made by the 'company' Harvard Communication; a Chinese idol girl group named SNH48 in a music video performing the same dances and wearing the same Japanese high-school uniforms as the Japanese idol girl group AKB48… These are just some objects on the list of *shanzhai* practices featured in this book's essays. By no means exhaustive, the list nonetheless shows the richness and diversity of the phenomenon as well as the difficulty in defining the concept; *shanzhai* has become a vernacular term that can refer to nearly anything, from consumer goods and cultural products, to even people.

The literal translation of *shan-zhai* 山寨 is 'mountain fortress', an image known from Shi Nai'an's famous novel *Water Margin* 水滸傳(1589), which tells a Robin Hood-like story about a group of 105 men and three women fighting a corrupt official in the Song Dynasty (960-1279). In the 1980s, the term *shanzhai* became mostly known for its reference to poor-quality, cheap counterfeits of brand products produced in the, often underground, factories in the Southern parts of China. In particular, the city of Shenzhen, which after its designation as one of the first Special Economic Zones (SEZ) in 1980 grew from a market town with 30,000 residents to a city of twelve million today, has become known as the centre of *shanzhai* production.

At the beginning of this century, *shanzhai* was mostly associated with copycat mobile phones made by small, semi-legal workshops, which were sold for around one third of the price of the original brands. Since both the producers and the consumers of these cheap mobile phones are migrant workers, *shanzhai* has come to be viewed as a product by and for the working class. While advanced-economy countries harshly criticise the (potential) violation of international Intellectual Property Laws (IPR) (copyrights, trademarks, and patents), many Chinese celebrate the 'democratising' power of *shanzhai* products and, in particular, of *shanzhai* mobile phones, providing access to the 'information have-less'.[32] According to a Chinese survey conducted in 2009, almost 65 per cent of the people 'supports' *shanzhai* as an expression of the 'voice of the people' (民间声音), while only 30 per cent are 'against' it and view it as 'counterfeit, piracy, and fraud'.[33] It is in this

32 Fan Yang, 'China's "Fake" Apple Store: Branded Space, Intellectual Property and the Global Culture Industry', *Theory, Culture & Society* 31.4 (2014): 71-96; Homi Bhabha, *The Location of Culture* (London: Routledge, 1994); Elizabeth Vann, 'The Limits of Authenticity in Vietnamese Consumer Markets', *American Anthropologist* 108.2 (2006): 286-95.

33 Five per cent had no opinion. Y. Hu, '*Shanzhai* de wenben jiexi yu dangdai meijie biaoyi [A textual analysis of *shanzhai* and its representation in modern media]', *Sanxia Luntan* 3.238 (2011): 10.

vein that some Chinese scholars regard *shanzhai* as a modern-day version of Maoist ideology, i.e. the Mass Line, the People's War, and the principle of self-reliance. Xiao Yuefan, for example, argues in this volume that Maoism 'has decisively pre-configured the contours within which present-day *shanzhai* activities operate'. Made by and for the masses, *shanzhai* products are the outcome of a collective practice without an identifiable individual designer, inventor, patent owner, or an official brand.

This collective creative process makes the phenomenon at odds with common notions of creativity as a form of individual and original expression. Then again, the practice of copying existing, 'original', brands is at odds with common notions of creativity itself, where authenticity and originality are generally juxtaposed with fakeness and copying. *Shanzhai* challenges this idea of authenticity as original expression in relation to IPR laws, a specific Euro-American concern, as also argued by Vann: IPR laws 'are employed to maintain clear distinctions between the authentic and the fake, the genuine and the counterfeit. International IPR laws [...] derive from a history of legal decisions in Europe and the United States that linked authorship to ownership, privileged originals over copies, and turned ideas into property'.[34] Likewise, scholars like Vann (2006), Bhabha (1994), and Taussig (1993) have pointed out how the concept of authenticity has often served as an exclusionary category in colonial settings in which the coloniser was always represented as 'the original' and indigenous cultural practices and products as 'inferior copies'. The act of *shanzhai*-ing thus disrupts taken-for-granted understandings of 'original/authentic' and 'copy/fake' that haunt the discourse of creativity by revealing their roots in a global legal-economic system of exclusion and inequality.

In this light, it is interesting to take a closer look at the most well-known *shanzhai* product: 'fake' mobile phones. While the first *shanzhai* mobile phones date from before the development of smartphones, they were among the first mobile phones to experiment with additional apps that would later be standard on smartphones. This had everything to do with the fact that these mobile phones were primarily consumed by the 'information have-less' who had little to no access to other technology and fully relied on their *shanzhai* mobile phones.[35] In sharp contrast to the earlier discussed Maoist readings of *shanzhai*, this innovative potential of *shanzhai* is frequently

34 Vann, 'The Limits of Authenticity', 288. See also Jeroen de Kloet and Chow Yiu Fai, '*Shanzhai* Culture, Dafen Art and Copyrights'. In *Routledge Handbook of East Asian Popular Culture*, eds. Chris Berry, Koichi Iwabuchi, and Eva Tsai (New York; London: Routledge, 2017), 229-41.

35 Fan Yang, 'China's "Fake" Apple Store', 86.

linked to the concept of 'creative destruction' (Schumpeter) in the domestic discourse; the capitalist idea that competition leads to innovation. Within the still strictly regulated Chinese economy that mainly focuses on Made in China, *shanzhai* becomes the place where innovation happens, where one can find the most creative people, products, and ideas, where the step to Created in China is taking place.

Importantly, *shanzhai* as a reproductive innovation process primarily reproduces *foreign* products *locally*. For this reason, it is also seen as a counterculture towards the Western 'global innovation culture' and the capitalist system by copying Western luxury goods into cheap and low-quality local products.[36] The idea of Chinese producers creatively 'destroying' Western brands is also behind the nationalist pride that many Chinese feel towards *shanzhai* products. 'Long live the Chinese people / Resolutely attack legitimate products, support pirate products', in the words of the Chinese netizen quoted at the opening of this section. One can even claim a quite specific history of copying here, Pang Laikwan argues that practices of copying defined life in China during the Cultural Revolution: 'The mastery of an art often begins with copying, and it is only through copying that the underlying rules and patterns can be revealed, and the new discovered. Copying is always the foundation of different art trainings. But the acts of copying exercised by the Chinese people went far beyond the confines of the arts'.[37]

At the same time, *shanzhai* can also turn *against* the Chinese state. 'Essentially, *shanzhai*, or copycatting practice, occurs when people cannot consume certain products because of either price or policy', Anthony Fung and Yiyi Yin observe in this volume. Just like there is more space for innovation and creativity in the production process of *shanzhai* luxury goods, the production of *shanzhai*-ed popular culture has opened up unique spaces for critical voices. Since 2008, the so-called Year of *Shanzha*i, there has been an explosion of *shanzhai* as, what Fan Yang calls 'counterfeit culture'.[38] The fast development of new media and technology have made a kind of 'do-it-yourself'-*shanzhai* possible, where websites like *Bilibili* and *Youku* are flooded with parody movies, music videos, and TV shows. For example, in his famous essay 'Copycat', the writer Yu Hua describes a *shanzhai* TV show

36 Josephine Ho, '*ShanZhai*: Economic/Cultural Production through the Cracks of Globalization', plenary presentation, ACS Crossroads: 2010 Cultural Studies Conference, Hong Kong, June 2010.
37 Pang, *The Art of Cloning – Creative Production During China's Cultural Revolution* (London: Verso, 2017), 247.
38 Fan Yang, *Faked in China: Nation Branding, Counterfeit Culture, and Globalization* (Indianapolis, IN: Indiana University Press, 2015).

made in the period of the Chinese milk powder scandal, 'in the ponderous tones of Network News they announced that the regular anchors had been poisoned by contaminated milk and rushed off to intensive care; they had been brought in at the last minute to deliver that evening's broadcast'.[39] Probably the most well-known example is the dozens of *shanzhai* Spring Festival Galas, some of which have attracted around two million viewers. While this becomes far less of an impressive number if one realises that the official Spring Festival Gala attracts around 800 million viewers, these spoof shows display a uniquely direct and open critique to the immensely popular and propagandistic TV show. This is particularly remarkable considering that while political satire, a genre more common in Western popular culture, is intensely censored in the Chinese context, framing it as a *shanzhai* practice seems to offer a discursive shield against official surveillance and enhance its survival and circulation. 'Seen in this way, it represents a challenge of the grassroots to the elite, of the popular to the official, of the weak to the strong', as Yu Hua asserts.[40] It is with these counterfeit cultural products, evoking a sense of illegality and subversion, that *shanzhai* regains the rebellious connotation from its etymological origin as the 'mountain fortress'.

The rebellious connotation will become all the more evident when we juxtapose the term *shanzhai* with other possible ways of naming something faking, copying, or counterfeiting. In the Chinese nomenclature, *jia* 假 (and its variations such as *jiamao* 假冒 and *zuojia* 做假), meaning fake, usually refers to things that are not the real ones, but pretend to be. It is sometimes applied to *shanzhai* products, and what is worth noting is that when the products or practices become harmful, they are invariably called *jia*. Reverting to Yu Hua's essay, the TV show is *shanzhai*, but the adulterated milk powder that became a health hazard to Chinese children will always be called *jia*. The latest scandal surrounding substandard vaccines shows similar linguistic use; they are *jia*, not *shanzhai*. This apparently impromptu vernacular distinction underwrites the positive connotation of *shanzhai*, of its alliance with the people, understood in juxtaposition to the state and corporate capital. However, just as the boundaries between *shanzhai* and *jia* are unsteady and porous, the rebellious potentials of faking, copying, and counterfeiting practices are always contingent. And it is this contingency and the potentiality opened by *shanzhai* practices that the inquiries in this book seek to examine and understand.

39 Yu Hua, *China in Ten Words*, trans. Allan H. Barr (New York: Pantheon Books, 2011), 186.
40 Ibid., 188.

As the examples above show, *shanzhai* is about copying, but not quite, as there is so much more to it than just copying. Following this insertion of contingency and potentiality in connecting *shanzhai* with creativity, one could reconceptualise it as a form of cultural translation: with each translation from an assumed original to a copy, meanings slip away while other meanings proliferate.[41] It is precisely in this process of translation that creativity comes into play. In their essay, Anthony Fung and Yiyi Yin analyse Chinese 'second-degree reproductions' of foreign cultural products such as Korean TV shows and Japanese anime, to argue that these radically *shanzhaied* versions 'emphasize a kind of autonomy and self-owned authority' that enable Chinese youth to create an alternative discourse within 'the complex dynamic between commercial culture and the nation's anti-globalisation policy'. In his study of Hong Kong-based artist Leung Mee-ping's work 'Made in Hong Kong/Shenzhen', Louis Ho challenges the 'individual model of creativity' that undervalues craft labour, asking 'who is the real artist of the work?' Drawing on Maoism, Xiao Yuefan brings in a very different conceptual framework for understanding *shanzhai* and traces the participatory aspirations of *shanzhai* in Mao's emphasis on art of and for the masses that served as a mobilisation vehicle for broader social equality. Lena Scheen questions why *shanzhai* products make us laugh and traces this apparently unsettling effect of *shanzhai* products in their 'resistance of the newness dogma dominating Euro-American cultures'. After providing a rich overview of the *shanzhai* phenomenon in all its various forms and readings, Stefan Landsberger suggests copyrighting the term *shanzhai* itself and to transform it into a brand that signifies the turn from 'made in China' to 'created by China'. Feng Fan reports on an exchange programme between Tsinghua University and the University of New York at Buffalo, where students made projects inspired by the question of whether the Chinese Dream is, in fact, a *shanzhai* version of the American Dream. In his field notes from interviews with television makers at Changsha's Hunan Satellite TV, Arjen Nauta explores how the notion of 'banal creativity' can help us understand how television makers conceptualise and interpret creativity. Finally, two Shenzhen-based artists and curators Dai Dai and Deng Chunru discuss their experiences of the 'two sides of SZ' – Shenzhen and *shanzhai* – in relation to their own works.

41 Jeroen de Kloet and Lena Scheen, 'Pudong: The *Shanzhai* Global City', *European Journal of Cultural Studies* 16.6 (2013): 706.

Digitisation

> Dragging around folders, zooming in on random objects and double-clicking around in open space. With computers omnipresent in both work and private lives, idle fiddling via digital interface has become the equivalent of doodling in notebooks. Enter Chinese artist Lin Ke – pronounced similar to 'link' – the master of digital procrastination.
>
> *Art magazine* Sleek[42]

How to think the third theme, the digital, in relation to the first two themes of this volume, boredom and *shanzhai*? The digital artist Lin Ke explicitly connects the themes of boredom, copy culture, and digitisation in his work, turning digital procrastination into works of art.[43] By blending layers of images taken from the internet with material from computer software and videos of himself, any meaningful distinction between what is 'real' or 'virtual', 'authentic' or 'copied' dissolves, reducing the spectator – those who are pleaded to 'like' him – to passive consumers of yet another stream of images. According to the above-cited art magazine *Sleek*, 'Ke describes the oversaturation-induced apathy that has left an entire generation of Internet users unenthusiastically switching between tabs for hours on end'.[44] Lin Ke aspired to become a painter, but when he could not afford a studio, and witnessing people around him wasting time behind computer and telephone screens, he decided to turn the computer into his studio and become a digital artist. Ironically, while immersing himself in digital cultures, he moved away from the city and is now living at the outskirts of Beijing, avoiding art openings and leading a quiet and solitary life. In his video work 'Like Me', Ke raps the foreshadowing postmodern words from a *Star Trek* episode from the 1960s, commenting on our increasingly digitised world, 'When dreams become more important than reality, you give up travel, building, creating. You just sit, living and re-living other lives left behind in the thought wreck'.

Since the emergence of the internet, there has been much discourse about its liberating potential for China.[45] Simultaneously, the Chinese internet

42 Donna Schons, 'The Chinese Artist Turning Digital Procrastination into Art', *SLEEK 54* (blog), accessed July 13, 2017, http://www.sleek-mag.com/2017/07/13/lin-ke-art/.

43 Ibid.

44 Ibid.

45 Yang Guobin, *The Power of the Internet in China: Citizen Activism Online*, reprint edition (New York: Columbia University Press, 2009).

is perceived as a controlling machine, warded off by what is termed the Great Firewall – a problematic, simplifying, and Orientalising metaphor as Lokman Tsui has argued convincingly.[46] Much research has problematised univocal and technological deterministic readings of the internet in China, arguing, for example, how its development is very much shaped by the rise of a new 'rule by morality' that is only understood from its specific socio-cultural context.[47] Others have shown how beyond the battle between utopian and dystopian narratives, something quite different seems to have emerged in China over the past decade: a fragmented digital realm that penetrates everyday life deeply and that has by now become a profoundly commercialised space. In the words of Jens Damm, 'This commercialisation of the Internet is significant because it has underpinned much of the trivialisation and de-politicisation of Internet content and usage in China. And in many ways, it is precisely this depoliticisation that has helped to create social spaces in which marginal groups such as gays and lesbians can claim a voice'.[48]

Much has changed over the last decade, aside from an intensified commercialisation, we can see the shift from computers towards mobile phones and the related emergence of platforms like *Taobao*, *Weibo*, Taxi Didi, Mobike and, the multi-functional app without which one cannot survive in China today, WeChat. We are now witnessing the platformisation of Chinese society in which social life is profoundly mediated and structured by digital technologies. While in the recent past such platforms were often celebrated as vehicles of the 'participatory society' and the 'sharing economy',[49] they often prove less progressive than they appear at first sight. Rather than simply stimulating citizen participation and entrepreneurialism, they enable the 'datafication' and 'commodification' of all social relations: collecting, algorithmically processing, circulating, and selling user data.[50] Whereas in

46 Lokman Tsui, 'An Inadequate Metaphor: The Great Firewall and Chinese Internet Censorship', *Global Dialogue* 9.1/2 (2007): 60-8.

47 Wai-chi Rodney Chu and Chung-tai Cheng, 'Cultural Convulsions: Examining the Chineseness of Cyber China'. In *Online Society in China: Creating, Celebrating, and Instrumentalising the Online Carnival*, eds. David Kurt Herold and Peter Marolt (London and New York: Routledge, 2011), 23-39.

48 Jens Damm, 'The Internet and the Fragmentation of Chinese Society', *Critical Asian Studies* 39.2 (2007): 291.

49 Henry Jenkins, *Convergence Culture: Where Old and New Media Collide* (New York: New York University Press, 2006).

50 José van Dijck and Thomas Poell, 'Understanding Social Media Logic', SSRN Scholarly Paper ID 2309065, Rochester, NY: Social Science Research Network, 2013, accessed July 13, 2017, https://papers.ssrn.com/abstract=2309065.

the West, governments are increasingly trying to regulate platforms like Uber and Airbnb, the case of China is quite peculiar given the already strong involvement of the government. The government is eager to embrace the political and economic potentials new technologies have to offer, earlier in its e-government initiatives, today in its Internet Plus policies. These are geared towards updating the conventional industries in China through the use of the internet and other information technologies. As Michael Keane shows in his chapter, related terms like the 'Cloud' and the 'Internet of Things' are all part of these new policies and resonate with today's global movement towards more surveillance and an intensified datafication of social life.

However, just like the *shanzhai* examples of counterfeit culture described in the previous section, digital technologies also allow for the constant making and circulation of creative spoofs and parodies, and in more general terms, they *afford* the proliferation of different creativities. One example is the contrasting of pictures of Xi Jinping and Obama with Winnie the Pooh that went viral in China, after which the government quickly censored any mention of the yellow bear. Such memes constantly stir up laughter as well as debate, and can be read as a visual guerrilla warfare revolving around pleasure and humour, rather than direct political critique.[51] It is not just the political context that may set digital cultures in China apart from other contexts, also the speed with which these cultures change, as well as their easy and rather smooth diffusion in particular towards the urban areas of China, seems to be extraordinary. Within less than a year, many young people in Beijing started to cycle again due to the massive distribution of rental bikes that are operated via an app (e.g. Ofo and Mobike). Food delivery to home turns out to be a way not just to save time, but also to avoid the air pollution and traffic jams. Due to the massive popularity of online shopping, streets are now occupied by delivery vans and motorbikes.

The Xi Jinping as Winnie the Pooh meme can be considered a bottom-up, everyday creative practice influenced by popular culture and facilitated by digital technologies; arguably a form of vernacular creativity. This term gestures to the mundane and everyday, to the banal and the trivial and, as such, to the constant making and circulation of GIF stickers on WeChat, the use of emoticons to replace words, and the memes discussed earlier. For Jean Burgess, 'one of the most useful questions cultural studies can ask about

51 See Bingchun Meng, 'From Steamed Bun to Grass Mud Horse: E Gao as Alternative Political Discourse on the Chinese Internet', *Global Media and Communication* 7.1 (2011):33-51; Thomas Poell, Jeroen de Kloet, and Guohua Zeng, 'Will the Real Weibo Please Stand up? Chinese Online Contention and Actor-Network Theory', *Chinese Journal of Communication* 7.1 (2014): 1-18.

new media is: "which technologies, practices and forms most effectively communicate vernacular creativity?"'[52] The word 'communicate' seems to be rather odd; instead, we wonder which practices *enable* or, following Keane in this volume, *afford* vernacular creativity. In Burgess's reading, these creativities increasingly turn consumers into producers, a reading we doubt given that only few users of, for example, YouTube, actually create videos themselves.[53] What the term 'vernacular creativity' does do, in our view, is to recuperate the banal and everyday and thus allow for an understanding of creativity that steers away from the celebration of the purely unique and the absolute original. Instead, we are left with new configurations of already existing materials, of *shanzhai*'ed versions of the real, and it seems to be that especially new media cultures allow for such reconfigurations. This resonates with Burgess's definition of creativity, which, for her, refers to 'the process by which available cultural resources (including both "material" sources – content, and immaterial resources – genre conventions, shared knowledges) are recombined in novel ways, so that they are both recognisable because of their familiar elements, and create affective impact through the innovative process of this recombination'.[54]

Aside from potentially being a boredom machine, new technologies also allow for the endless replication, repetition, and circulation of images, sounds, and words. As such, they somehow resemble a *shanzhai*'ing machine. The artwork *Chinternet Plus – A Counterfeit Ideology* by Miao Ying is a parodic take on the CCP's Internet Plus policies (integration of the internet with traditional industries to fuel economic growth). The work is exhibited, but also exists online.[55] It was featured in *The New Normal – China, Art and 2017* exhibition in the Ullens Center in Beijing's 798 Art Zone in the Spring of 2017. There, screenshots and videos were displayed and played, remaining partly hidden as they were surrounded by a wall punctured with holes, clearly a reference to the Great Firewall. The celebratory discourse on creativity is pushed to the parodic extreme in this work, for example when claiming that 'the most creative aspect of Chinternet Plus is that it makes the original a counterfeit. In this reality, commentary does not just replace the content, here reality serves comments and gets tips'. The digital

52 Jean Burgess, 'Hearing Ordinary Voices: Cultural Studies, Vernacular Creativity and Digital Storytelling', *Continuum* 20.2 (2006): 207.

53 José van Dijck, *The Culture of Connectivity: A Critical History of Social Media*, 1st edition (New York: Oxford University Press, 2013).

54 Burgess, 'Hearing Ordinary Voices', 206.

55 Available at https://www.chinternetplus.com, accessed July 13, 2017.

and the copy conflate in the work, thus instructions are given on how to counterfeit a logo:

> To create a counterfeit logo, one has to walk a fine balance to make it just right. The logo cannot look too much like the original, yet still should remind people of the original. One needs to play it cool; in order to do so, one has to catch the 'spirit' of the original.

An app is presented that will turn the polluted sky blue, on the website and in the clips, accompanied by a bizarre mixture of cultural texts from around the world, presenting a cosmopolitan bricolage of people, places, and artefacts. 'Reality is not efficient enough, but your smart apps are', the artist (prophetically?) proclaims.

In his essay, Michael Keane examines traditional roots of Chinese creativity in Confucian and Daoist approaches and its connection to nature and the market, showing how the concept of affordances provides a new way of understanding Chinese creativity. The essays by Zeng Guohua and Chen Siyu both present an analysis of the rapid emergence of digital cultures in their discussions of the omnipresence of WeChat in payment and retail. Zeng analyses the popularity for online shopping for safe, natural food, as a response to issues of food unsafety, whereas Chen probes the intricacies of online payment, connecting this to both issues of control and regulation as well as to enabling forms of vernacular creativity.

Zoénie Liwen Deng shows in her field notes that parody and critique are not the only mode Chinese artists mobilise when engaging with new technologies; the art projects *Cyber Nails* deliberately mingle the online world of WeChat with offline practices, exploring the schism, or what Deng terms the Mobius-strip-like relationship, between online and offline, between showing and seeing. Rowan Parry reports from his research on independent documentaries how new technologies, in particular the internet and WeChat, are profoundly changing film cultures in China not only at the textual level, but also in terms of production, circulation, and reception. In their dialogue with Lo Yin Shan, curators Isaac Leung and Janet Fong explore the changes in both society and the art world enabled by new technologies. Like most authors, while acknowledging the transforming potential of platforms like WeChat, they are highly suspicious of their possible drawbacks; there seem to be far too many minuses to the Internet Plus.

By bringing the notion of creativity into conversation with the themes of boredom, *shanzhai*, and digitisation – i.e. notions that are usually juxtaposed with creativity but that are common features of creative practices in China

– this collection offers alternative pathways to rethink creativity. It may help to look for new ideas, which may not be new after all; nevertheless, isn't this *shanzhai*-ing of knowledge, of creative practices, the mechanism, the process that propels the continuous proliferation of new thoughts, of new practices?

Bibliography

Anderson, Ben. 'Time-stilled Space-slowed: How Boredom Matters', *Geoforum* 35.6 (2004): 739-54.

Bhabha, Homi. *The Location of Culture* (London: Routledge, 1994).

Burgess, Jean. 'Hearing Ordinary Voices: Cultural Studies, Vernacular Creativity and Digital Storytelling', *Continuum* 20.2 (2006): 201-14.

Chu, Wai-chi Rodney, and Chung-tai Cheng. 'Cultural Convulsions: Examining the Chineseness of Cyber China'. In *Online Society in China: Creating, Celebrating, and Instrumentalising the Online Carnival*, eds. David Kurt Herold and Peter Marolt (London and New York: Routledge, 2011), 23-39.

Dalle Pezze, Barbara, and Carlo Salzani. 'Introduction: The Delicate Monster: Modernity and Boredom'. In *Essays on Boredom and Modernity,* eds. Barbara Dalle Pezze and Carlo Salzani (Amsterdam: Rodopi, 2008), 5-34.

Damm, Jens. 'The Internet and the Fragmentation of Chinese Society', *Critical Asian Studies* 39.2 (2007): 273-94.

de Kloet, Jeroen, and Lena Scheen. 'Pudong: The *Shanzhai* Global City', *European Journal of Cultural Studies* 16.6 (2013): 692-709.

de Kloet, Jeroen, and Anthony Y.H. Fung. *Youth Cultures in China* (Malden, MA: Polity, 2017).

de Kloet, Jeroen, and Chow Yiu Fai. '*Shanzhai* Culture, Dafen Art and Copyrights'. In *Routledge Handbook of East Asian Popular Culture,* eds. Chris Berry, Koichi Iwabuchi, and Eva Tsai (New York; London: Routledge, 2017), 229-41.

Gardiner, Michael E. 'Henri Lefebvre and the "Sociology of Boredom"', *Theory, Culture & Society* 29.2 (2012): 37-62.

Goodstein, Elizabeth S. *Experience Without Qualities: Boredom and Modernity* (Stanford, CA: Stanford University Press, 2008).

Groys, Boris. *On the New.* Translated by G. M. Goshgarian (Brooklyn, NY: Verso, 2014).

Ho, Josephine. '*ShanZhai*: Economic/Cultural Production through the Cracks of Globalization'. Plenary presentation, ACS Crossroads: 2010 Cultural Studies Conference, Hong Kong, June 2010.

Hoffman, Eva. *How to be Bored* (London: The School of Life, 2016).

Hu, Y. '*Shanzhai* de wenben jiexi yu dangdai meijie biaoyi [A textual analysis of *shanzhai* and its representation in modern media]', *Sanxia Luntan* 3.238 (2011): 9-14.

Jenkins, Henry. *Convergence Culture: Where Old and New Media Collide* (New York: New York University Press, 2006).
Judkins, Rod. *The Art of Creative Thinking: 89 Ways to See Things Differently* (New York: TarcherPerigee, 2016).
Jullien, Francois. *The Silent Transformation.* Translation by K. Fijalkowski and Michael Richardson (London: Seagull Books, 2011).
Keane, Michael. *Creative Industries in China: Art, Design and Media* (Malden, MA: Polity, 2013).
—. *Created in China: The Great New Leap Forward* (Abingdon, NY: Routledge, 2007).
Kierkegaard, Søren. *Either/Or: A Fragment of Life* (Copenhagen: University Bookshop Reitzel, 1843).
Lefebvre, Henri. 'Toward a Leftist Cultural Politics: Remarks Occasioned by the Centenary of Marx's Death'. In *Marxism and the Interpretation of Culture*, eds. L. Grossberg and C. Nelson (Urbana, IL: University of Illinois Press, 1988), 75-88.
Lorey, Isabell, with a foreword by Judith Butler. *State of Insecurity: Government of the Precarious.* Translation by Aileen Derieg. Reprint edition (London; New York: Verso, 2015).
McRobbie, Angela. *Be Creative: Making a Living in the New Culture Industries* (Cambridge, UK; Malden, MA: Polity, 2016).
Meng, Bingchun. 'From Steamed Bun to Grass Mud Horse: E Gao as Alternative Political Discourse on the Chinese Internet', *Global Media and Communication* 7.1 (2011): 33-51.
Miao, Ying. *Chinternet Plus – A Counterfeit Ideology.* Accessed July 13, 2017. https://www.chinternetplus.com.
Musharbash, Yasmine. 'Boredom, Time, and Modernity: An Example from Aboriginal Australia', *American Anthropologist* 109.2 (2007): 307-17.
O'Connor, Brian. *Idleness – A Philosophical Essay* (Princeton, NJ: Princeton University Press, 2018).
Pang, Lai-kwan. *The Art of Cloning – Creative Production During China's Cultural Revolution* (London: Verso, 2017).
—. '"China Who Makes and Fakes": A Semiotics of the Counterfeit', *Theory, Culture & Society* 25.6 (2008): 117-40.
Poell, Thomas, Jeroen de Kloet, and Guohua Zeng. 'Will the Real Weibo Please Stand up? Chinese Online Contention and Actor-Network Theory', *Chinese Journal of Communication* 7.1 (2014): 1-18.
Postman, Neil. *Amusing Ourselves to Death: Public Discourse in the Age of Show Business.* 20th Anniversary Edition (London: Penguin Books, 2005 [1985]).
Rofel, Lisa. *Desiring China: Experiments in Neoliberalism, Sexuality, and Public Culture* (Durham, NC: Duke University Press, 2007).

Schons, Donna. 'The Chinese Artist Turning Digital Procrastination into Art', *SLEEK 54* (blog). Accessed July 13, 2017. http://www.sleek-mag.com/2017/07/13/lin-ke-art/.
Spacks, Patricia Meyer. *Boredom: The Literary History of a State of Mind* (Chicago, IL: University of Chicago Press, 1995).
Svendsen, Lars. *A Philosophy of Boredom.* Translation by John Irons (London: Reaktion, 2005).
Taussig, Michael. *Mimesis and Alterity: A Particular History of the Senses* (London: Routledge, 1993).
Toohey, Peter. *Boredom: A Lively History* (New Haven, CT: Yale University Press, 2012).
Tsui, Lokman. 'An Inadequate Metaphor: The Great Firewall and Chinese Internet Censorship', *Global Dialogue* 9.1/2 (2007): 60-8.
van Dijck, José. *The Culture of Connectivity: A Critical History of Social Media.* 1st edition (New York: Oxford University Press, 2013).
van Dijck, José, and Thomas Poell. 'Understanding Social Media Logic'. SSRN Scholarly Paper ID 2309065. Rochester, NY: Social Science Research Network, 2013. Accessed July 13, 2017. https://papers.ssrn.com/abstract=2309065.
Vann, Elizabeth. 'The Limits of Authenticity in Vietnamese Consumer Markets', *American Anthropologist* 108.2 (2006): 286-95.
Yang, Fan. *Faked in China: Nation Branding, Counterfeit Culture, and Globalization* (Indianapolis, IN: Indiana University Press, 2015).
—. 'China's "Fake" Apple Store: Branded Space, Intellectual Property and the Global Culture Industry', *Theory, Culture & Society* 31.4 (2014): 71-96.
Yang, Guobin. *The Power of the Internet in China: Citizen Activism Online.* Reprint edition (New York: Columbia University Press, 2009).
Yu, Hua. *China in Ten Words.* Translation by Allan H. Barr (New York: Pantheon Books, 2011).

About the Authors

Jeroen de Kloet is Professor of Globalisation Studies and Director of the Amsterdam Centre for Globalisation Studies (ACGS) at the University of Amsterdam, and affiliated to the School of Music and Recording Arts, Communication University of China. In 2010, he published *China with a Cut – Globalisation, Urban Youth and Popular Music* (Amsterdam University Press). He wrote, together with Chow Yiu Fai, *Sonic Multiplicities: Hong Kong Pop and the Global Circulation of Sound and Image* (Intellect, 2013) and edited together with Lena Scheen *Spectacle and the City – Chinese Urbanities in Art and Popular Culture* (Amsterdam UP, 2013). He wrote, together with Anthony Fung, *Youth Cultures in China* (Polity, 2017). See also www.jeroendekloet.nl.

Chow Yiu Fai is Associate Professor at the Department of Humanities and Creative Writing of Hong Kong Baptist University. His current research projects concern creative labour, creative practices, and single women. His publications include *Caring in Times of Precarity* (Palgrave, 2019) and *Sonic Multiplicities* (co-authored, Intellect, 2013). Next to his academic life, Chow is an award-winning creative writer in lyrics and prose. Lately, Chow has been increasingly involved in multi-media and visual art projects.

Lena Scheen is Assistant Professor of Global China Studies at NYU Shanghai, and Global Network Assistant Professor at NYU. Scheen's research explores the social and cultural impact of China's fast urbanisation, focusing on Shanghai and storytelling. Her publications include the monograph *Shanghai Literary Imaginings: A City in Transformation* (AUP, 2015) and the edited volume, together with Jeroen de Kloet, *Spectacle and the City: Chinese Urbanities in Popular Art and Culture* (AUP, 2013). She is currently working on her new manuscript *Shanghai Storytelling: City Myths, Historical Memory, and Place Attachment.*

II 'Creative China' and Its Potential to Problematise Western-Modern-Romantic Ideologies of Creativity

Eitan Wilf

Abstract

Recurrent, most recently Western-Modern-Romantic, ideologies conceptualise creativity as the solitary, ex nihilo creation of products of self-evident and universal value by highly exceptional individuals. These ideologies have had a tremendous impact beyond the context and time of which they were the product. They have naturalised a number of binary oppositions such as creativity/imitation, creativity/routinisation, and creativity/reproduction, often superimposed on other binary oppositions such as modernity/tradition and the west/the rest, respectively. These ideologies have thus contributed to the production and reproduction of different forms of social hierarchy, and also hindered a deeper understanding of creative agency. Against this backdrop, this chapter theorises the potential of boredom, imitation and digitisation to problematise Western-Modern-Romantic discourses and practices of creativity.

Keywords: creativity, ideology, social hierarchy, cultural variety

Introduction

Not being a sinologist, I am unqualified to write anything substantial that can be anchored in the ethnographic context of creative practices in contemporary China. What I can do is to point at the tremendous potential of the three discussion axes identified in the book's introduction to productively problematise Western-Modern-Romantic discourses and practices

---

Kloet, Jeroen de, Chow Yiu Fai, and Lena Scheen (eds), *Boredom,* Shanzhai, *and Digitisation in the Time of Creative China*. Taylor & Francis Group, 2019
DOI: 10.5117/9789462984745_WILF

of creativity, which have had a widespread impact on the ways in which creativity has been understood and practiced in and beyond the West. I will argue that each of the three discussion axes – boredom, imitation, and digitisation – undermines a different building block of Romantic ideologies of creativity and that, in doing so, these axes can help us to problematise the very notion of creativity against the backdrop of its ideological simplification and reification, which have functioned in the production and reproduction of different forms of social hierarchy.

Boredom

To tease out the potential of the first discussion axis, boredom, we can begin by noting not only the fact that creativity in its Romantic form entails the very opposite of boredom, but also the intellectual history of such entailment, which goes back to ancient Greece. Ancient Greek theories of poetic inspiration attributed it to the divine intervention of entities such as the muses.[1] 'Divine inspiration' was one of the key explanations for poetic creativity first formulated by the ancient Greeks. The theory of divine inspiration, which stipulated that 'the poem is dictated to the poet by a visitor from without',[2] accounted for the following four features of poetic practice, according to ancient Greek commentators:

> (a) The composition is sudden, effortless, and unanticipated. The poem or passage springs to completion all at once, without the prior intention of the poet, and without that process of considering, rejecting, and selecting alternatives which ordinarily intervenes between the intention and the achievement. (b) The composition is involuntary and automatic; it comes and goes at its own pleasure, independently of the will of the poet. (c) In the course of composition, the poet feels intense excitement, usually described as a state of elation and rapture, but occasionally said to be racking and painful in its initial stages, though followed by a sense of blissful quiescence. (d) The completed work is as unfamiliar and surprising to the poet as though it had been written by someone else.[3]

1 Meyer H. Abrams, *The Mirror and the Lamp* (Cambridge, MA: Harvard University Press, 1971), 189-90.
2 Ibid., 190.
3 Ibid., 189.

This theory of divine inspiration, although slightly out of favour in the context of increased rationalism in early modern Europe, returned with full force at the end of the eighteenth century with the rise of Romanticism but with one significant change: inspiration was interiorised in the person via organic metaphors of the independent inner gestation and growth of the genius's inner nature. According to Romantic ideologies of poetic inspiration, 'an inspired poem or painting is sudden, effortless, and complete, not because it is a gift from without, but because it grows of itself, within a region of the mind which is inaccessible either to awareness or control'.[4]

Note the ways in which all four characteristics of this notion of poetic inspiration entail different experiential dimensions that are the exact opposite of boredom. The first characteristic stipulates the lack of any prior planning and editing considerations in the creation of a work of art, i.e. the lack of the often tedious 'process of considering, rejecting, and selecting alternatives'. The second characteristic describes a form of practice that is the exact opposite of the nine-to-five job format: poetic inspiration presumably 'comes and goes at its own pleasure', it cannot be tamed or routinised. It thus maintains an element of surprise that is antithetical to the boredom that often accompanies routinisation. The third characteristic explicitly stipulates that the experience of being poetically inspired is antithetical to boredom: the experience is that of 'elation and rapture'. Finally, the fourth characteristic, similarly to the second one, introduces surprise and novelty into the life of the poet in that the work of art produced as a result of being inspired appears 'unfamiliar and surprising' to the poet. These notions of creativity thus represent a concerted effort to erase boredom as an element of creativity.

However, contrary to these notions, boredom is inherent to creative practice. First, boredom is inseparable from the often long and tedious process of acquiring the skills and techniques that are the building blocks of almost any practice; not only of seemingly basic 'techniques of the body' such as swimming or digging,[5] but also of creative practices more specifically, whether these practices entail learning to play a specific instrument, draw properly, mix paint to achieve a desired hue, execute a specific dancing move, and so forth. A theory of creativity predicated on divine inspiration erases the skill and technique that are the conditions of possibility of creative practice, and consequently the boredom and tedious repetition involved in the process of acquiring them. A focus on the role of boredom in, and on

4 Ibid., 192; Charles Taylor, *Sources of the Self: The Making of the Modern Identity* (Cambridge, MA: Harvard University Press, 1989).

5 Marcel Mauss, 'Techniques of the Body', *Economy and Society* 2.1 (1973): 70-88.

the more mundane and less colourful dimensions of, creative practice thus properly situates this practice in relation to its actual conditions of possibility and existence and, more generally, in relation to social life in general of which boredom, routinisation, and tedious repetition are part and parcel. In short, it humanises it against the backdrop of its ideological deification.

Equally important is the fact that boredom, when theorised as an inseparable dimension of creativity, has the potential to undermine different forms of social hierarchy predicated on the prevalent ideologies of creativity I described above for, both in its ancient Greek and Romantic versions, poetic inspiration is based in the assumption that the poet or artist is an elected, exceptional creature, one whose gifts are bestowed upon him either from without (by a divine entity), or from within (by way of natural genius). It is indicative that the centrality of the notion of divine inspiration resulted in the fact that, save for a few artistic movements, training has played a minor role in the understanding of artistic creativity in the West.[6] Noteworthy are 'the absence of any references in early Greek literature to the possibility of one poet teaching another',[7] the hostility in the Renaissance to all but the most elementary forms of teaching and the preference given instead to 'God's grace' as the basis of great art,[8] and the full-blown Romantic sanctification of 'artistic freedom' at the expense of codified rules, even in the context of institutionalised art education in the European academies.[9] Such frameworks have thus instituted a form of social hierarchy in which some people are naturally gifted and some are not and where training plays little to no part at all. This form of social hierarchy exists on a higher level of social reality, too, for example, in the dichotomy between the West and the rest, which has been superimposed on and supported by the dichotomy between creativity and mere technique or skill, respectively. In the context of some creative practices such as jazz music, the West has often supported its own image as creative by projecting to Asian cultures the stereotype of soulless technique.[10]

6 Eitan Wilf, 'Streamlining the Muse: Creative Agency and the Reconfiguration of Charismatic Education as Professional Training in Israeli Poetry Writing Workshops', *Ethos* 41.2 (2013): 127-49.

7 Penelope Murray, 'Poetic Genius and its Classical Origins'. In *Genius: The History of an Idea*, ed. Penelope Murray (Oxford: Blackwell, 1989), 11; Ray McDermott, 'Materials for a Confrontation with Genius as a Personal Identity', *Ethos* 32.2 (2004): 278-88.

8 Carl Goldstein, *Teaching Art: Academies and Schools from Vasari to Albers* (Cambridge: Cambridge University Press, 1996), 25.

9 Arthur Efland, *A History of Art Education: Intellectual and Social Currents in Teaching the Visual Arts* (New York: Teachers College Press, 1990), 52.

10 Taylor Atkins, *Blue Nippon: Authenticating Jazz in Japan* (Durham, NC: Duke University Press, 2001).

A focus on boredom as an essential element of creative practice undermines these and other forms of social hierarchy. First, if creative practice is based in the often boring and tedious acquisition of skills and technique, it means that creativity is not the exclusive and rarefied form of practice it has seemed to be, but rather something that many people can potentially master if they are given the proper conditions to do so. Second, the realisation that skills and technique are the bedrock of creative practice points at the economic capital needed to acquire them, and thus at economic disparity as a significant determinant of what seems to be the exclusive and rarefied nature of creativity. This is essentially an argument about the ways in which economic capital is converted into cultural capital that is misrecognized as natural, rather than as the product of such conversion.[11] Third, anchoring creative practice in the more routinised and boring dimensions of social life points at its conventionality or arbitrariness as opposed to its universally necessary nature as stipulated by theories of divine inspiration or natural genius. Creative practice, including the perception of what counts as the successful products of this practice, become the product of relatively arbitrary conventions or routines organised in culturally and socially specific ways, akin to what the sociologist Howard Becker has called 'art worlds'.[12]

Imitation

To tease out the potential of the second discussion axis, imitation, we should similarly begin by noting not only the fact that creativity in its Romantic form entails the very opposite of imitation, but also the intellectual history and social determinants of such entailment. In the context of one of Western modernity's key narratives about itself, the relationship between imitation and creativity reverberates with the question of the relationship between personal autonomy and tradition.[13] It thus concerns core Enlightenment ideas such as freedom from tradition as directly linked to the progress of humankind. For instance, at one point in his *Critique of the Power of Judgment*, Emmanuel Kant argues that 'the product of a genius [...] is an example, not for imitation (for then that which is genius in it and constitutes the spirit of the work would be lost), but for emulation by another genius,

11 Pierre Bourdieu, *Distinction* (London: Routledge, 1981).

12 Howard Becker, *Art Worlds* (Berkeley, CA: University of California Press, 1982).

13 Wilf, 'Rituals of Creativity: Tradition, Modernity, and the "Acoustic Unconscious" in a U.S. Collegiate Jazz Music Program', *American Anthropologist* 114.1 (2012): 32-44.

who is thereby awakened to the feeling of his own originality, to exercise freedom from coercion in his art'.[14] Thus, Kant understands the individual's ability to exercise his or her creative faculties as a practice of freedom, a manifestation of enlightened agency, which imitation can only corrupt. This intellectual legacy has had a definitive impact on the ways in which modernity has been imagined. Creativity has come to index modernity whereas imitation has come to connote tradition.

However, imitation is inherent to creative practice, and vice versa. First and foremost, if, as I have mentioned above, any creative practice necessitates the learning of skills and technique, such learning is overwhelmingly accomplished through imitation.[15] Equally important is the fact that to effectively, i.e. creatively, break from an already established way of doing things novices must first learn the ins and outs of already established conventions. In other words, they must learn to be competent participants in an existing art world so that they can identify and capitalise on opportunities to do things differently. Such learning is overwhelmingly achieved via imitation.

Furthermore, if creativity always already requires some form of imitation, imitation is always already creative in some way. Recent anthropological studies have emphasised the creativity that imitation necessitates to be successful:

> Copying or imitation, we argue, is not the simple, mechanical process of replication that it is often taken to be, of running off duplicates from a template, but entails a complex and ongoing alignment of observation of the model with action in the world. In this alignment lies the work of improvisation. The formal resemblance between the copy and the model is an outcome of this process, not given in advance. It is a horizon of attainment [...].[16]

Imitating, following a model, and engaging in forms of highly regimented, rule-governed behaviour socialise participants into creativity if only because they must constantly adapt their real-time activity of imitating to the general patterns and the general patterns to the contingency of the unique situation of imitating.

14 Immanuel Kant, *Critique of the Power of Judgment* (Cambridge: Cambridge University Press, 2008 [1790]), 195-96.

15 Mauss, 'Techniques of the Body'.

16 Tim Ingold and Elizabeth Hallam, 'Creativity and Cultural Improvisation: An Introduction', in *Creativity and Cultural Improvisation,* eds. Elizabeth Hallam and Tim Ingold (Oxford: Berg, 2007), 5. Emphasis in original.

Finally, a number of anthropologists working in ethnographic contexts typically associated with rote memorisation and ritualised forms of imitation have problematised the Romantic distinction between, on the one hand, creative, authentic, or spontaneous interiority and, on the other hand, external, mediated conventions.[17] They have problematised this distinction by arguing that individuals in these ethnographic contexts hone such interiority by following, imitating, and inhabiting the external, conventional, and highly regimented forms and behaviours associated with it – a process whose dialectical nature anthropologists have tried to capture by coining phrases such as 'rehearsed spontaneity'[18] and 'rituals of creativity'.[19] In so doing, they have uncovered the ways in which the Western-Modern-Romantic notion of the singularity and hence the inimitability of creative practice and its products results from a specific temporal framework within which time is understood to be moving in a linear fashion and past events cannot be relived and inhabited in the present by radical imitation. Acknowledging the cultural specificity of this temporal framework and the existence of highly different temporal frameworks in which past events – including creative ones – can be relived and inhabited in the present via radical forms of imitation offers another vantage point from which to undermine the dichotomy between creativity and imitation.

A thorough analysis of the Western-Modern-Romantic dichotomy of creativity and imitation also suggests that it is overdetermined by the rise of possessive individualism as a modern normative ideal and the centrality of property rights as a defining feature of the modern autonomous individual.[20] Individual creativity has come to be perceived as a prototypical form of an inalienable possession.[21] Tropes of possessive individualism pervaded early Romantic notions of organic creativity. Thus, in an essay that had a tremendous impact on Romanticism, entitled 'Conjectures on Original Composition', Edward Young expressed his disdain for learning by arguing that 'learning is borrowed knowledge; genius is knowledge innate, and quite

17 Wilf, 'Semiotic Dimensions of Creativity', *Annual Review of Anthropology* 43 (2014): 397-412.

18 Saba Mahmood, 'Rehearsed Spontaneity and the Conventionality of Ritual: Disciplines of Salat', *American Ethnologist* 28.4 (2001): 827-53.

19 Wilf, 'Rituals of Creativity'.

20 James Leach, 'Creativity, Subjectivity, and the Dynamic of Possessive Individualism', in *Creativity and Cultural Improvisation*, eds. Elizabeth Hallam and Tim Ingold (Oxford: Berg, 2007), 99-116.

21 Anette Weiner, *Inalienable Possessions: The Paradox of Keeping While Giving* (Berkeley, CA: University of California Press, 1992).

our own'.[22] For Young, creativity cannot be based on any form of debt qua imitation.

What does a focus on imitation as an inseparable dimension of creativity risk reveal? First, it highlights the fact that creative practice is always already found in a web of social relations and historical transactions and debt. The Western-Modern-Romantic notion of the singularity of creativity and of its products amounts to nothing less than what Marx called 'commodity fetishism'.[23] Marx argued that consumers are unaware of the fact that when they buy a certain commodity they enter into relations with the workers who produced it, i.e. that they pay for the work of others. Consumers think that they pay a price that is inherent to the commodity and that is essentially tied to its 'nature', as if the commodity came ready-made and had a nature of its own. Thus, from products of a brutal economic system these commodities are transformed into objects endowed with quasi mystical, sui generis qualities. We can similarly argue that the Western-Modern-Romantic notion of creative practice is responsible for 'creativity fetishism', i.e. the fetishisation of the artist and of his or her products, in the context of which they are conceptualised as autochthonous entities that bear no relation to other entities and that are endowed with their own self-anchoring value. Such a framework hides the different forms of debt that make creative practice possible. Conversely, a focus on the role of imitation in creative practice, similar to the focus on boredom, rehumanises creative practice by placing it squarely within the web of social relations and transactions that are its conditions of possibility.

Digitisation

To tease out the potential of the third discussion axis, digitisation, we should similarly begin by noting not only the fact that creativity in its Romantic form entails the very opposite of digitisation, but also the intellectual history and social determinants of such entailment. A fruitful way of approaching this topic is provided by Walter Benjamin's seminal essay, 'The Work of Art in the Age of Its Mechanical Reproduction'.[24] The first thing to note is that in this essay Benjamin argues that the value of the artwork, which he calls

22 Edward Young, *Conjectures on Original Composition* (London: Miller, 1759), 36.

23 Karl Marx, 'Capital, Volume One'. In *The Marx-Engels Reader,* ed. Robert C. Tucker (New York; London: Norton, 1974), 319-29.

24 Walter Benjamin, 'The Work of Art in the Age of Mechanical Reproduction', in *Illuminations* (New York: Schocken, 1969), 217-51.

its 'aura', results from the artwork's radical analogue character. One key dimension in Benjamin's definition of the artwork is 'uniqueness':

> Even the most perfect reproduction of a work of art is lacking in one element: its presence in time and place, its unique existence at the place where it happens to be. [...] The presence of the original is the prerequisite to the concept of authenticity. [...] The whole sphere of authenticity is outside of technical – and, of course, not only technical – reproducibility.[25]

Benjamin argues that the artwork's value, its aura, is in part the result of the artwork's physical singularity, the fact that there is only one such object in the world. The artwork is a specific and unique material object, distinguished by virtue of its radical concreteness and analogue individuation:

> This unique existence of the work of art determined the history to which it was subject throughout the time of this existence. This includes the changes which it may have suffered in physical condition over the years as well as the various changes in its ownership. The traces of the first can be revealed only by chemical or physical analysis which it is impossible to perform on a reproduction; changes of ownership are subject to a tradition which must be traced from the situation of the original.[26]

Although Benjamin seems to acknowledge that the artwork's analogue character is ultimately bound up with processes of authentication such as chemical analysis and historical research, he seems to suggest that these processes are strategies capable of actual verification.

However, anthropologists have argued that the value associated with the artwork's presumed analogue character is produced in specific baptismal and performative events that are the subject of subsequent citations, resulting in indexical chains that people are socialised to recognise as indexing authenticity and aesthetic necessity.[27] For example, the careful presentation of an artwork's past provenance in an art auction is a type of performative authenticating speech that ritually transforms the artwork into a singular exemplar. The same goes for the artwork's ritualised chemical analysis. The singularity of an object of creativity, which amounts to an insistence on its radical analogue character, is thus the product of emergent communicative

25 Ibid., 222.
26 Ibid.
27 Wilf, 'Semiotic Dimensions of Creativity'.

events during which its history is carefully entextualised, i.e. given coherence that detaches it from the object's past messy circulation and dubious singular material qualities.[28] The result of such performative events is an inversion whereby an object is understood to have a distinguished biography and material characteristics because of its singular value, whereas in reality it often comes to have this value because it has been imbued with such a biography and material characteristics by means of ritualised and performative world-making forms of authentication.

Understanding the socially fabricated nature of the identification between the products of creative practice and total analogue individuation has the potential to undermine different forms of social hierarchy predicated on such identification. Note that the second dimension of Benjamin's definition of the value of the artwork has to do with the fact that its physical uniqueness and total individuation result in its radical distance and hence inaccessibility.[29] The total individuation of the artwork, the fact that there exists only one such physical object, guarantees a form of social hierarchy between those people who can gain access to the artwork and those people who cannot.

The realisation that works of art need not be defined by some kind of radical analogue nature because the latter is the product of performative processes of authentication means that digitisation, the penultimate form of creative production and reproduction in the contemporary moment, need not be understood as antithetical to creative practice, but rather as a technological shift that can enrich it – a point that Benjamin was aware of, of course, as is evident in his analysis of the revolutionary potential of photography and film (albeit these technologies were not digital at the time he wrote his essay). Digitisation primarily provides an infrastructure of creative practice that democratises, to some extent, the production and consumption of art by virtue of the ubiquity of digital technologies and of their ease of use, as well as by their tendency to afford a kind of transcendence of the here and now via the easy production of copies, editing, and transmission of information. Inasmuch as copying, editing, and flow of information have always existed to some extent, the unruly nature of digitisation does not amount to a truly revolutionary phase in the conditions of possibility of creative practice, but rather to the increased visibility and democratisation of practices that have until now been unacknowledged and the access to which has been granted to only a select few.

28 Greg Urban, 'Entextualization, Replication, and Power', in *Natural Histories of Discourse*, eds. Michael Silverstein and Greg Urban (Chicago, IL: University of Chicago Press, 1996).
29 Benjamin, 'The Work of Art', 224.

Conclusion

'Creative China', inasmuch as it is predicated on the explicit embrace rather than suppression of boredom, imitation, and digitisation, offers a unique opportunity for theorists and practitioners to problematise reified notions of creative practice and the different forms of social hierarchy that they have helped to naturalise on different levels of social reality, including the levels of individuals, institutions, ethnic groups, and nation states. The stakes are broader than those that concern creative practice, for creative practice is a core dimension of human agency in general. A reconceptualisation of our understanding of creative practice, then, has the potential to lead to a reconceptualisation of our understanding of human agency.

Bibliography

Abrams, M. H. *The Mrror and thc Lamp* (Cambridge, MA: Harvard University Press, 1971).

Atkins, Taylor. *Blue Nippon: Authenticating Jazz in Japan* (Durham, NC: Duke University Press, 2001).

Becker, Howard. *Art Worlds* (Berkeley, CA: University of California Press, 1982).

Benjamin, Walter. 'The Work of Art in the Age of Mechanical Reproduction'. In *Illuminations* (New York: Schocken, 1969), 217-51.

Bourdieu, Pierre. *Distinction* (London: Routledge, 1981).

Efland, Arthur. *A History of Art Education: Intellectual and Social Currents in Teaching the Visual Arts* (New York: Teachers College Press, 1990).

Goldstein, Carl. *Teaching Art: Academies and Schools from Vasari to Albers* (Cambridge: Cambridge University Press, 1996).

Ingold, Tim, and Elizabeth Hallam. 'Creativity and Cultural Improvisation: An Introduction'. In *Creativity and Cultural Improvisation,* eds. Elizabeth Hallam and Tim Ingold (Oxford: Berg, 2007), 1-24.

Kant, Immanuel. *Critique of the Power of Judgment* (Cambridge: Cambridge University Press, 2008 [1790]).

Leach, James. 'Creativity, Subjectivity, and the Dynamic of Possessive Individualism'. In *Creativity and Cultural Improvisation,* eds. Elizabeth Hallam and Tim Ingold (Oxford: Berg, 2007), 99-116.

Mahmood, Saba. 'Rehearsed Spontaneity and the Conventionality of Ritual: Disciplines of Salat', *American Ethnologist* 28.4 (2001): 827-53.

Marx, Karl. 'Capital, Volume One'. In *The Marx-Engels Reader,* ed. Robert C. Tucker (New York; London: Norton, 1974), 294-442.

Mauss, Marcel. 'Techniques of the Body', *Economy and Society* 2.1 (1973): 70-88.

McDermott, Ray. 'Materials for a Confrontation with Genius as a Personal Identity', *Ethos* 32.2 (2004): 278-88.

Murray, Penelope. 'Poetic Genius and its Classical Origins'. In *Genius: The History of an Idea,* ed. Penelope Murray (Oxford: Blackwell, 1989), 1-13.

Taylor, Charles. *Sources of the Self: The Making of the Modern Identity* (Cambridge, MA: Harvard University Press, 1989).

Urban, Greg. 'Entextualization, Replication, and Power'. In *Natural Histories of Discourse,* eds. Michael Silverstein and Greg Urban (Chicago, IL: University of Chicago Press, 1996).

Weiner, Anette. *Inalienable Possessions: The Paradox of Keeping While Giving* (Berkeley, CA: University of California Press, 1992).

Wilf, Eitan. 'Semiotic Dimensions of Creativity', *Annual Review of Anthropology* 43 (2014): 397-412.

—. 'Streamlining the Muse: Creative Agency and the Reconfiguration of Charismatic Education as Professional Training in Israeli Poetry Writing Workshops', *Ethos* 41.2 (2013): 127-49.

—. 'Rituals of Creativity: Tradition, Modernity, and the "Acoustic Unconscious" in a U.S. Collegiate Jazz Music Program', *American Anthropologist* 114.1 (2012): 32-44.

Woodmansee, Martha. 'Genius and the Copyright'. In *The Author, Art, and the Market: Rereading the History of Aesthetics* (New York: Columbia University Press. 1996), 35-56.

Young, Edward. *Conjectures on Original Composition* (London: Miller, 1759).

About the Author

Eitan Wilf is an Associate professor in the Department of Sociology and Anthropology at the Hebrew University of Jerusalem. He received his PhD in anthropology from the University of Chicago in 2010. His research interests focus on the institutional transformations of creative practice in the United States. He has conducted ethnographic research on the academisation of art-training, the development of art-producing computerised algorithms and sociable robots, and the cultural logic of business innovation. He is the author of *School for Cool: The Academic Jazz Program and the Paradox of Institutionalized Creativity* (University of Chicago Press, 2014), and *Creativity on Demand: The Dilemmas of Innovation in an Accelerated Age* (University of Chicago Press, 2019).

Section 1

Boredom

Essays

1.1 Create No More!

Clutter and Boredom, a Hong Kong Perspective

Anneke Coppoolse

Abstract
In Hong Kong, waste and order have, since the colonial days, been inextricably related. "Things" (commodities, everyday objects) were welcomed and detested. Many were of importance to international trade, but some were viciously eliminated by the authorities. Today, like in any modern city, the consumption of things still purposes modern life while the underlying problem of 'boredom' assures continuous ridding and cluttering of others. Artists display alternative ways to embrace boredom and clutter. In Hong Kong, the oeuvres of two prominent 'collecting artists', Kwok Mang-ho (Frog King) and the late Ha Bik-chuen, help reinterpret what could be called an 'aesthetics of overabundance' as they feed off, speak to, and move away from the consumer culture that occupies modern life.

Keywords: aesthetics, art, boredom, clutter, Hong Kong, modernity

> [...] there is in the life of a collector a dialectical tension between the poles of disorder and order.[1]

Henri Lefebvre argues that the relation between interest and boredom may well be 'one of the dialectical movements within "modernity", one

1 Walter Benjamin, 'Unpacking My Library: A Talk about Book Collecting', in *Illuminations*, ed. Hannah Arendt (New York: Schocken Books, 1968), 60.

Kloet, Jeroen de, Chow Yiu Fai, and Lena Scheen (eds), *Boredom,* Shanzhai*, and Digitisation in the Time of Creative China*. Taylor & Francis Group, 2019
DOI: 10.5117/9789462984745_CH1.1

of its concealed movements'.[2] 'The interesting' or 'the new' is short-lived: it wears out and becomes boring.[3] Modern boredom induces a desire for stimulation, which is often found in commodities. Modern life can therefore be seen as structured following a constant rejection of commodities and a demand for more. This is a global situation that prompts particular stories in different localities: also, in Hong Kong where space is limited and everyday conduct has been historically informed by an insistence on order and cleanliness as instructed via laws and cultures of governance. Hong Kong is at the same time a place where consumption forms the base for urban living; a place defined by free market politics and transnational trade; a place where the production and consumption of things (commodities) is part of its global appeal. Modern life in Hong Kong can therefore be understood to be negotiated between things, their rapid redundancy, and desired order.

As the urge for the interesting is generated by a permanent sense of boredom, unrestrained consumerism and frenzied pursuit of the new are popular ways of coping,[4] which – in turn – is a direct effect of the altered experience of time in modern life. With the mechanical beat that ignited the capitalist mode of production, 'human activity came to be subjected to sameness and repetition'[5] and the linearity of this 'commodity time'[6] induced an urge for stimulation as found in the continuous supply of new things. The lingering of stuff in modern spaces can, thus, be understood as a result of the modern rhythm of life.

The boredom that this new rhythm of life brings about, however, is only temporarily soothed with novelty.[7] As Salzani asserts, although the newness of things stimulates demand, mass production is 'the eternal return of the same': 'the experience is new yet the shock with which it occurs repeats itself. Novelty – its impact on modern life – is generated by 'the "fata-morgana" logic of commodity production',[8] which causes overstimulation and no

2 Henri Lefebvre, *Introduction to Modernity: Twelve Preludes, September 1959-May 1961* (London: Verso, 1995), 260.

3 Michael Gardiner, 'Henri Lefebvre and the "Sociology of Boredom"', *Theory, Culture & Society* 29.2 (2012): 49.

4 Ibid.

5 Barbara Pezze and Carlo Salzani, 'Introduction', in *Essays on Boredom and Modernity,* eds. Barbara Pezze and Carlo Salzani (Amsterdam: Rodopi, 2009), 13.

6 Lefebvre, *Critique of Everyday Life. Vol. II* (London: Verso, 2002), 49.

7 Pezze and Salzani, 'Introduction', 13.

8 Carlo Salzani, 'The Atrophy of Experience: Walter Benjamin and Boredom', in *Essays on Boredom and Modernity,* eds. Barbara Pezze and Carlo Salzani (Amsterdam: Rodopi, 2009): 134.

time to digest. In this process, proper experience 'wastes away': amidst the overload of things, aesthetic experience[9] becomes 'neurasthenic'.[10]

Yet, Walter Benjamin sees also value in boredom: 'it can form the beginning of an awareness that the monotony of the present will only end with a resolution of the deeper contradictions of society, and the creation of an alternative society based on true creativity and pleasure'.[11] As boredom allows awareness of the undesirability of monotonous life, there is opportunity for alternative experience. Although consumer activities do not involve creativity as such, the everyday is not only defined by repetition.[12]

To Lefebvre, everyday life is both 'humble and sordid', and rich in potential; it is the 'space-time of voluntary programmed self-regulation', but also of utopian possibility.[13] Such possibility can be found in acts of tinkering.[14] Tinkering is a way of 'making do': an 'art of living'.[15] Further, following Jacques Rancière's theory of the 'redistribution of the sensible',[16] the kind of utopian possibility Lefebvre describes eventuates in the ways certain artists deal with the material consequence of modern boredom: clutter.

Not only experience is 'decaying' as a symptom of (material) overstimulation and a lack of time. As Benjamin argues that the altered experience in modern life is directly related to changes in 'concrete societal artefacts',[17] also these artefacts waste away. 'Things', piling up at unprecedented speeds, following the continuous production of the new and prompted by boredom and the logic of commodities, are instantly wasting away. Yet, even though clutter may be one of the most apparent material indicators of the modern human malaise, it also appears to be the material *par excellence* to contribute to the imagining of more colourful futures. Since the avant-garde, it often contributed to, '[t]he creation of artworks, philosophy or even entire cities

9 Aesthetic experience here meaning 'a shared corporeal disclosure of the sensible world through which notions of meaning and value are formulated' (Lefebvre, 1995, p. 165).

10 Benjamin, *The Writer of Modern Life: Essays on Baudelaire* (Cambridge: The Belknap Press, 2006), 133.

11 Joe Moran, 'Benjamin and Boredom', *Critical Quarterly* 45.1-2 (2003): 180.

12 Gardiner, 'Henri Lefebvre', 48.

13 Lefebvre, *Critique of Everyday Life*, 72.

14 Michel de Certeau, *The Practice of Everyday Life* (Berkeley, CA: University of California Press, 1984), xv.

15 Michel de Certeau, Luce Giard and Pierre Mayol, *The Practice of Everyday Life, Vol. 2: Living and Cooking* (Minneapolis, MN: University of Minnesota Press, 1998), xiv.

16 Jacques Rancière, *The Politics of Aesthetics: The Distribution of the Sensible* (London: Bloomsbury Academic, 2004).

17 Michael Jennings, 'Introduction', in *The Writer of Modern Life: Essays on Baudelaire*, ed. Walter Benjamin (Cambridge: The Belknap Press, 2006), 8.

[...] through the free (and indeed playful) appropriation of space and time, which is starkly opposed to the purely "technical mastery of material nature which produces products and exchange values"'.[18] It has contributed to what Lefebvre calls *oeuvres*.

Indeed, art has engaged itself with the materiality of modernity for more than a century. From modernism onwards, it has questioned and revisited the ordering system of modernity – through political statements or alternative imaginations about waste and clutter: from Charles Baudelaire's *Le Vin de Chiffoniers* (The Ragpicker's Wine) to Chinese contemporary artist Song Dong's highly personal installation 'Waste Not' – a perfectly organised exhibition of clutter from his mother's home.

What interests me is not so much the redundant thing as reconsidered in art. Instead, I look at ways in which lingering stuff – in the context of modern Hong Kong – is negotiated in art. This chapter therefore first lays out the context of Hong Kong as a place where order and tidiness have been strived for since its establishment; it then elaborates an understanding of clutter in relation to modernity; after which it considers clutter and chaotic display as recurring themes in works of art. It then zeros in on Hong Kong art, detailing how clutter and unwanted things are key to the oeuvres of two of Hong Kong's most renowned collecting artists, Ha Bik-chuen and Kwok Mang-ho (Frog King), artists whose oeuvres also extend stories about modern art in Hong Kong and the influences of the West and local histories, more broadly; the chapter concludes with a remark on the possibility of an aesthetics of overabundance in an embrace of the monotony that capitalist production engendered.

On Order and Unwanted Things: Hong Kong

In 1844 – only a few years after the establishment of Hong Kong – order and unwanted matter became legally connected with the introduction of the *No. 5 Good Order and Cleanliness Ordinance*.[19] The Ordinance was part of a colonisation scheme that echoed both British law and its culture of colonisation.[20] The scheme at large subjected the Chinese people in Hong

18 Lefebvre, *Writings on Cities* (Oxford: Basil Blackwell, 1996), 175-76.

19 The Colony of Hong Kong, *Ordinance No. 5: Good Order and Cleanliness* (1844), 14, accessed August 23, 2016, http://oelawhk.lib.hku.hk/archive/files/2b4c38ded8e441159b4dcb8f05f1073d.pdf.

20 David Donald, *A Financial Centre for Two Empires: Hong Kong's Corporate, Securities and Tax Laws in Its Transition from Britain to China* (Cambridge: Cambridge University Press, 2014), 25-26.

Kong to a regime of control while Ordinance No. 5 specifically formalised the conversion of the organisation of everyday conduct and concerns about cleanliness as it combines a prohibition against the throwing of rubbish with bans on undesired forms of public behaviour.[21]

Demands on sanitation and behavioural control became specifically apparent when the bubonic plague broke out in 1894. The colonial government had wrongly determined that its cause was a 'lack of hygiene' among what was referred to as the 'Chinese underclass'.[22] The plague was seen as a '"filth" disease' that appeared only in places where poor people lived.[23] As a result, the authorities sanitised Chinese people's homes and ordered strict segregation between Europeans and the rest of the population of Hong Kong. This proves, besides seriously misguided disease aetiology and the prejudiced perceptions of Chinese habits[24] also, as Robert Peckham recently argued, the misconception that 'Chinese things' were 'sources of likely contagion'.[25]

The sanitation campaign involved total elimination of Chinese residents' properties. Furniture, clothes – 'things' referred to as 'rubbish'[26] – were piled up onto the streets for permanent destruction. What is more, as Peckham also suggested, the plague came to be seen as an apparent 'Chinese thing'. The microscope had rendered the bacterium of the plague visible which had made it besides symptomatically particular, also particularly 'material'.[27] Indeed, although 'Chinese things' were key to Britain's wealth, they became increasingly perceived of as suspicious. Even opium, although grown in India and mainly exported by Britain to China, came to be viewed as a 'pernicious Chinese commodity': 'a plague'.[28]

Control over the everyday conduct of Chinese members of Hong Kong's society was, during the years of the plague, forcefully executed while not only habits were misinterpreted, disease and commodities, things and trash, belongings and bacteria, were confused towards the general conception that every*thing* Chinese was unhygienic. Yet, the unsanitary

21 The Colony of Hong Kong, *Ordinance No. 5*, 14.

22 Carol Benedict, *Bubonic Plague in Nineteenth-Century China* (Stanford, CA: Stanford University Press, 1996), 143.

23 Ibid.

24 Ibid.

25 Robert Peckham, 'Hong Kong Junk: Plague and the Economy of Chinese Things', *Bulletin of the History of Medicine* 90.1 (2016): 1.

26 'The Plague in Hong Kong', *Times*, August 28, 1894, p. 6.

27 Peckham, 'Hong Kong Junk', 17.

28 Ibid., 5.

state of Hong Kong was already addressed in government documents as early as 1854. Colonial surgeon Dr. J. Carroll Dempster had specifically written about Tai Ping Shan (a market area on the slope of Victoria Peak) where two large open drains 'receive[d] all the refuse of the district through which they pass'.[29] He added, '[t]he great want of privies and suitable depots for dirt is observable everywhere the native population reside'.[30]

The Sanitary Board was only established in the late ninetieth century, scrutinising housing conditions in Chinese neighbourhoods, and considered drainage and sewer systems. Yet, it was the plague that prompted action. Segregation of the population and related urban expansion allowed new Chinese housing with improved sanitary facilities.[31] Disease, however, kept spreading as the population grew – specifically cholera and smallpox. The colonial government maintained segregation to protect the health of Europeans, caring little about general public health.[32]

The expanding of the city went on until the late 1930s[33] and commenced again after the Japanese occupation (1941-1945) as the population continued to grow. With the arrival of new immigrants from China, Hong Kong's manufacturing industry bolstered while specifically in the 1950s and 1960s also the production of plastic toys, watch parts, and electronics boomed.[34] The related 'gathering momentum of social development' fostered what in government documentation is described as 'a sense of civic pride' among the people of Hong Kong (Information Services Department),[35] which then led to a friendlier approach to demands on cleanliness through public participation campaigns in the 1960s and 1970s.

The Lap Sap Chung (垃圾蟲, 'Litterbug') campaigns (1970s) were most telling. Litterbug – a green monster – was the mascot of a range of campaigns

29 Drainage Services Department, *Sewerage and Flood Protection: Drainage Services 1841-2008* (Hong Kong: Drainage Services Department, The Government of Hong Kong Special Administrative Region, 2008), 10, accessed July 15, 2016, https://www.dsd.gov.hk/EN/Files/publications_publicity/Publications/2008book.pdf.

30 Ibid.

31 Ibid.

32 Ka-che Yip, 'Segregation, Isolation, and Quarantine: Protecting Hong Kong from Diseases in the Pre-war Period', *Journal of Comparative Asian Development* 11.1 (2012): 107-8.

33 Ibid., 29-36.

34 Hong Kong Memory, 'Industrialization in Post-war Hong Kong', *Hong Kong Memory*, 2012, accessed January 4, 2017, http://www.hkmemory.hk/collections/postwar_industries/industrialization_in_postwar_hong_kong/index.html#&panel1-1.

35 Information Services Department, 'Of Green Monsters and Broken Daggers', *GIS Through the Years*, 1999, accessed August 2, 2016, http://www.info.gov.hk/isd/40th/13.html.

that promoted hygiene and public order.[36] Litterbug represented trash and bad littering habits, which threatened modern Hong Kong life. The campaign posters depicted, for instance, people chasing Litterbug over the beach, the monster luring over the Hong Kong skyline from behind, or cleaners facing him with brooms. This illustrated a binary opposition between society and waste, in keeping with the larger and still apparent modern obsession with order.[37]

The ordering of places in the name of purity is a modern idea.[38] However, modern practices of managing chaos produce a binary other, resulting in a world continuously confronted with more chaos.[39] As Dennis Smith explains it, 'the classifying broom does not sweep clean; it just moves the dust around'.[40]

Everywhere does the ridding of unwanted things purpose modern living, while – today – fast fashion, planned obsolescence,[41] and other systems that build rapid redundancy into the lifespans of products, assure simultaneous admission of others. Even sustainable innovation is increasingly absorbed into the system of consumption now that green products are fashionable.[42]

Overconsumption is a global issue[43] with local consequences. It is a problem that in different contexts presents both global conditions and particular stories. In Hong Kong – which is like any modern city negotiated through binary oppositions such as order/chaos, society/trash – the apparent disconnect between the drive to buy and the need for clean presents its own narrative: from concerns of the colonial authorities with Chinese things to the general conception that trash does not belong.

36 Ka-che Yip, 'Segregation', 29-36.

37 Zygmunt Bauman, *Modernity and Ambivalence* (Cambridge: Polity Press, 1991), 7.

38 Dipesh Chakrabarty, *Habitations of Modernity: Essays in the Wake of Subaltern Studies* (Chicago, IL: University of Chicago Press, 2002), 65-66.

39 Lawrence Grossberg, *Cultural Studies in the Future Tense* (Durham, NC: Duke University Press, 2010), 87.

40 Dennis Smith, *Zygmunt Bauman: Prophet of Postmodernity* (Cambridge: Polity Press, 1999), 147.

41 'Planned obsolescence' involves the production of commodities with the intention to have them quickly become useless so that repeated purchasing is (or seems) necessary (Bulow, p. 729).

42 Fiona Allon, 'Green Renovations, Eco-Homes and Sustainable Home Improvement'. In *Ethical Consumption: A Critical Introduction,* eds. Tania Lewis and Emily Potter (London; New York: Routledge, 2010), 205.

43 Thomas Princen, Michael Maniates, and Ken Conca, eds. *Confronting Consumption* (Cambridge, MA: MIT University Press, 2002), 2.

A Piling up of Things: Clutter

The Oxford Dictionary defines waste as, '[u]nwanted or unusable material, substances, or by-products'.[44] A.R. Ammons, in his book-length poem *Garbage*, calls the scale at which it exists 'the poem of our time[45]: it is material left for poets to 'revamp' for that it is like a heap of 'used-up language'.[46] Indeed, waste has at one moment or another been acknowledged as such. It has been done away with. Yet, between the buying of things and their making redundant, there is stuff that is neither one nor the other. Or both. It may be 'wasted' in the sense of 'used or expended carelessly or to no purpose'.[47] It may be things that have been trashed and, in another moment, carefully collected. It may be new. Forgotten.

Waste directly relates to 'the excesses of unregulated, dirty capitalist production; or the obscenity of overconsumption'.[48] It is presented as *the* by-product of consumption.[49] Yet, specifically when considering consumption in places where space is limited, it becomes apparent that said obscenity involves besides matter made redundant also piles of things (commodities or their residues) that have been stalled, awaiting further usage or dumping. These are things existing, often by chance or unintentionally, in a space between (or within) society and trash. My point is that there is stuff left to linger. Clutter. Anna McCarthy, however, addressing its digital variant, regards it a 'buzzing keyword' of the 21st century which rhetoric 'treats it as an aesthetic problem, a psychological symptom, and an economic liability plaguing the on- and offscreen worlds of contemporary subjects'.[50]

44 Oxford English Dictionary Online, s.v. 'Waste', accessed September 25, 2016, https://en.oxforddictionaries.com/definition/waste.

45 Archie Randolph Ammons, 'Garbage', in *Best of the Best American Poetry: 25th Anniversary Edition*, eds. David Lehman and Robert Pinsky (New York: Scribner Poetry, 2013): 7.

46 David Lehman, 'A. R. Ammons, The Art of Poetry No. 73', *The Paris Review* 139 (1996), accessed December 22, 2016, https://www.theparisreview.org/interviews/1394/a-r-ammons-the-art-of-poetry-no-73-a-r-ammons. This perspective also prompts Baudelaire's description of the ragpicker, which is to be understood as a metaphor 'for the poetic method' (Benjamin, 2006, p. 108).

47 Oxford English Dictionary Online, s.v. 'Waste'.

48 Gay Hawkins, *The Ethics of Waste: How We Relate to Rubbish* (Lanham, MD: Rowman and Littlefield Publishers, 2006), 6.

49 For example, Jennifer Clapp, 'The Distancing of Waste: Overconsumption in a Global Economy', in *Confronting Consumption*, eds. Thomas Princen, Michael Maniates, and Ken Conca (Cambridge, MA: The MIT Press, 2002), 155-76; Kevin Hetherington, 'Secondhandedness: Consumption, Disposal, and Absent Presence', *Environment and Planning D – Society and Space* 22 (2004): 157-73.

50 Anna McCarthy, 'Clutter and Clodder', *Social Text* 34.3 (128) (2016): 127.

Lingering stuff is not exactly waste in the sense of unwanted or unusable matter. It may still be a nuisance for that it is an 'untidy collection of things'.[51] Most importantly, clutter piles up, like waste, due to a continuous focus on and desire for new things, which goes against the modern imagination of tidy cities.[52] McCarthy asks herself why clutter must be controlled. She can only provide one answer: 'because it threatens to expose the disorder of our selves'.[53]

Unwanted Things, Collected in Art: Aesthetics

In the arts, the materiality of the everyday has received much attention in the past 150 years. Specifically, found objects and forms of waste have been adapted in works of art in diverse ways.[54] In modernist art and in work produced thereafter, random, everyday things were used to challenge bourgeois aesthetics in which art and life were seen as greatly distinct.[55] In other words, in an attack on the 'ordering-system of modernity', the triviality of everyday life was magnified, which also accentuated the problems of capitalism more generally.[56]

Avant-garde artists began using forms of collage and bricolage. Most of their work was (and is) not so much intended to provoke, but rather to understand and juxtapose artistic form and life. Specifically, since the mid-twentieth century, such assembled works of found objects and images helped to think through 'the nature of reality, the nature of painting itself, and the methods by which creative thought is organised'.[57] The avant-garde unleashed 'a new revolutionary understanding of reality'.[58]

In this line of thought, Jacques Rancière sees much of the tradition not only working towards new understanding but engaging new 'sensible forms and material structures for a life to come'.[59] As he is concerned with the distribution of the sensible and finds in certain forms of art modes of

51 Oxford English Dictionary Online, s.v. 'Clutter', accessed September 25, 2016, https://en.oxforddictionaries.com/definition/waste.

52 Hawkins, 'Waste in Sydney: Unwelcome Returns', *PMLA* 122.1 (Special Topic: Cities) (2007): 3348.

53 McCarthy, 'Clutter and Clodder', 128.

54 Gillian Whiteley, *Junk: Art and the Politics of Trash* (London: I. B. Tauris, 2010), 7.

55 Maurizia Boscagli, *Stuff Theory: Everyday Objects, Radical Materialism* (New York; London: Bloomsbury, 2014), 234.

56 Ibid.

57 The Museum of Modern Art, *Press Release: The Art of Assemblage* (New York: MoMA, 1961), 1.

58 Nikos Papastergiadis, *Spatial Aesthetics: Art, Place, and the Everyday* (Amsterdam: Institute of Network Cultures, 2010), 22.

59 Jacques Rancière, *Aisthesis: Scenes from the Aesthetic Regime of Art* (London: Verso, 2013), 22-23.

redistribution, the aesthetic avant-garde allows experience of a new kind of division between what is and what is not sensible,[60] specifically by creating new experiences through an assemblage and display of things and messages, even sounds and smells.

Also, in contemporary art, unwanted stuff and found objects keep recurring and assemblage and collection are preferred methods. As Gillian Whiteley states that, 'trash has become the trope of the turn of the 21st century',[61] Nicolas Bourriaud comments that, 'the market has become an omnipresent referent'.[62] He argues that the flea market is the dominant visual model: it is where old things converge and await new purpose[63] – possibly the place of clutter. Chaotic display is the favoured aesthetic across the contemporary art world.

One artist who follows this aesthetic is Benin-based Georges Adéagbo. He has been assembling found objects for decades, making installations of stuff from and about particular places. He often collects things at flea markets and does not consider himself to be making art installation but installations about art. Another such artist who adapts to the aesthetic of the market is Surasi Kusolwong, who was trained in both Germany and his home country Thailand. Kusolwong makes installations with cheap Thailand-produced and assembled products, often presenting them in the form of rubbish-like piles or markets.[64] American artist Jason Rhoades (1965-2006), finally, made installations that highlighted consumer culture, labour, and obsolescence.[65] Bouriaud describes Rhoades' work as a collection of objects expressed in one single form – that of a market – where the individual objects do not entirely sit together.[66]

Many of the works that adapt to chaotic display also reveal a specific relation to time: they either directly engage the mechanical beat of capitalist production, or they propose alternative rhythms. Visual artist Tim Gaudreau, for instance, photographically collected the material remainders of his own consumer habits in a yearlong project that he titled, 'Trash: 365 Days of Photographing Everything That I Throw Out' (2006). Justin Gignac's ongoing Garbage project equally adapted to commodity time, as it involves

60 Rancière, *The Politics of Aesthetics*, 88.

61 Whiteley, *Junk*, 8.

62 Nicolas Bourriaud, *Postproduction*, 2nd edition (New York: Ram Publications & Dist, 2005 [2002]), 28.

63 Ibid.

64 Bourriaud, *Postproduction*, 28-31.

65 Ingrid Schaffner, 'Jason Rhoades, Rour Roads: A Case Study of Contemporary Art and Conservation', *Voices of Contemporary Art Journal* (2015), accessed January 8, 2017, http://journal.voca.network/jason-rhoades-four-roads/.

66 Bourriaud, *Postproduction*, 27.

'hand-picked' garbage from New York City, presented in transparent boxes.[67] Also the earlier-mentioned installation 'Waste Not' by Chinese artist Song Dong displays time (and the mundane) as it involves things from his mother, acquired through decades-long collecting.

All of these works must still be understood in context – 'situated within specific cartographies, chronologies and ethnographies'.[68] Gignac's boxed garbage items, for instance, cannot be about anything other than the place in which they are made. Put differently, New York City is probably one of few places where the selling of its rubbish as art would be successful. Kusolwong's work is equally context-specific, Thailand and its products being at the heart of his works, the theme explores the country's cultural and economic position in a global context. Yet, regardless of the particular stories that come with all of these artists' works, they have found in the collecting, assembling, (even 'marketing') of things their preferred form.

Not all the works above can redistribute the sensible. Specifically, political art – relating to those artworks that plainly address the problem of the commodity – only participates in the already existing discourse:

> To ask, How can one escape the market? is one of those questions whose principal virtue is one's pleasure in declaring it insoluble. Money is necessary to make art [...] to make a living you have to sell the fruits of your labor. So, art is a market, and there's no getting around it. For artists as for everyone else, there's the problem of knowing where to plant one's feet, of knowing what one is doing in a particular place, in a particular system of exchange. One must find ways to create other places, or other uses for places. But one must extricate this project from the dramatic alternatives expressed in questions like, How do we escape the market, subvert it, etc.?[69]

Addressing the redistribution of the sensible, Rancière points out work that surpasses critical narratives. He does not think that the concept 'avant-garde' can do this per se. Yet, the kind of work the tradition produced often services the breaking away from mere critiquing, because it explores 'forms of life' – often through assemblage – while 'the fundamental question is to explore the possibility for play'.[70] It explores 'how to produce forms for the

67 NYC Garbage, 'The Story behind HYC Garbage', accessed October 3, 2016, http://nycgarbage.com/about/.

68 Whiteley, *Junk: Art and the Politics of Trash*, 12.

69 Rancière, 'Art of the Possible: Fulvia Carnevale and John Kelsey in Conversation with Jacques Rancière', *Artforum* 45.7 (2007): 263.

70 Ibid.

presentation of objects, forms for the organisation of spaces, that thwart expectations'.[71]

Kusolwong's installations – although they bluntly address consumerism – allow such playful exploration. In *Golden Ghost: The Future Belongs to Ghosts* (2011), he presented an enormous pile of redundant fabric with hidden in it golden necklaces. Inviting visitors to find these necklaces (for keeps) by climbing the pile and searching through the stacks of fabric, he implemented an aspect of play that transformed the capitalist time/money binary into a play/gift phenomenon.[72] Products of mechanical reproduction – those temporary fixes for boredom in the age of linear rhythm – are in his work transformed into objects of playful engagement.

This answers to what Rancière calls the 'aesthetic regime of art':[73] a regime that 'institutes the opening up of art onto the everyday and the ordinary [...] by its ability to produce a world held in common'.[74] The aesthetic regime of art revives the ordinary and it transforms our experience of it.[75]

The Artist as Collector: Ha Bik-chuen

In the context of Hong Kong, where, as in any modern city, a continuous negotiation between clutter and imagined order plays out in the spaces of the everyday, art presents its own narratives. There are two *oeuvres*[76] in particular that cannot be ignored when considering practices of collection and assemblage in relation to ideas of overabundance and sensible experience in Hong Kong.

The late Ha Bik-chuen (1925-2009) is possibly the most renowned collecting artist that the territory has known. Besides his remarkable assemblages also the way in which he came to his work can, to an extent, be seen as symptomatic of the Hong Kong of his time. When Ha came from the Chinese

71 Ibid.

72 The Museum of Modern Art, 'Surasi Kusolwong', 2015, accessed October 4, 2016. http://momaps1.org/exhibitions/view/345.

73 Rancière, *The Politics of Aesthetics.*

74 Ben Highmore, *Ordinary Lives: Studies in the Everyday* (London; New York: Routledge, 2011), 49-50.

75 Ibid., 51.

76 I mean oeuvre in the way Lefebvre's conceptualises it: as 'the creation of artworks, philosophy or even entire cities, along with their attendant use-values, through the free (and indeed playful) appropriation of space and time, which is starkly opposed to the purely "technical mastery of material nature which produces products and exchange values"' (Gardiner, p. 48).

mainland to Hong Kong in the late-1950s, he started making handicrafts to earn a living: 'he took orders from local factories for paper flowers and bamboo baskets while beginning to study art on his own'.[77] A self-taught artist with in his early days no means to buy art books, he would pass by street vendors in the evenings and pick out pages of old magazines that featured works of art and other images.[78] This is where his collecting started and he became influenced by artists such as Picasso, Miro, Klee, and Matisse.[79]

Alongside his craftwork and while studying art by himself, Ha also started making prints and later sculptures usually out of bamboo, iron and other scrap materials. With time, he managed to graduate from making crafts to making art, using any form of media that he could find:[80] 'raw materials for making art can be picked up everywhere'.[81] He once described his work as 'junks-cum-art' exercises[82] and is for this reason one of Hong Kong's most prominent bricolage artists, although I have never seen him described as such. A self-taught artist focusing on everyday objects as he created his work with a collection of found materials, Ha has 'made do' in his training and in his work.

Besides his artworks, it is specifically his life-long mode of collection and assemblage that speaks to the narrative about Hong Kong, clutter, and proper aesthetic experience. Instead of describing his individual works, I focus therefore on his art as a process. A master collector, he has throughout the years continued to collect anything art-related (books, catalogues, cut-outs). He has also, since the 1960s, photographed over 1.5 thousand exhibitions in Hong Kong and abroad, capturing both artworks and their audiences.[83] Many of his collected images were further adapted into collages and collections kept in boxes, files, and books. So, when he died in 2009, he left behind – in his rooftop studio and library space – an archive full of photographs, files, and art books which pages were patched with cut-outs.[84]

77 Ha Bik-chuen, *Ha Bik-chuen: Ink at 76 (Exhibition and Sale)* (Hong Kong: Grotto Fine Art Ltd, 2001).

78 Ibid.

79 Man-hung Chan, 'Ha Bik-chuen – Sentient Beings: Solo Exhibitions of Relief Prints', in *Catalogue of Solo Exhibition* (Hong Kong: China Oil Painting Gallery Ltd., 1997), 6-7.

80 Ibid.

81 Anne Lam, *Explorer – The Art of Ha Bik-chuen* (Hong Kong: Artrend, 2004), 10-11.

82 Ha Bik-chuen, *Explorer – The Art of Ha Bik-chuen* (Hong Kong: Artrend, 2004).

83 Bernice Chan, 'A New-Found World: Hong Kong Artist Ha Bik-chuen's Legacy', *South China Morning Post*, June 9, 2015, accessed September 24, 2016, http://www.scmp.com/lifestyle/arts-entertainment/article/1817056/new-found-world-hong-kong-artist-ha-bik-chuens-legacy.

84 Ibid.

1.1.1 Ha Bik-chuen studio

Courtesy of the Ha Family and Asia Art Archive
Photo: Jack Chueng of NOTRICH MEDIA, Hong Kong

His family recently donated his collection to Hong Kong-based Asia Art Archive (AAA). Soon after, AAA organised an exhibition of fractions of this collection, titled *Excessive Enthusiasm: Ha Bik Chuen and the Archive as Practice* (2015). Although it is expected to take years to conserve Ha's archive, the exhibition allowed a first glance at his persistent collecting.[85] AAA's online Ha Bik Chuen Archive, further, is slowly evolving into a database that does not just present Ha's collection work, but allows a view of the history of Hong Kong art, which is still far less understood than Western art history.[86] Ha's collage books and photographic contact sheets provide a personal perspective. At the same time, it is increasingly claimed that not only his sculptures, prints and paintings belong to his oeuvre, but that also his collection forms part of it: 'the archive as practice', as per AAA's exhibition title.

Ha Bik-chuen's initial need to 'make do' with whatever he could find upon his arrival to Hong Kong, is not unique to him alone. Many immigrants

85 Ibid.

86 Asia Art Archive, 'Ha Bik Chuen Archive', accessed September 24, 2016, http://www.aaa.org.hk/Collection/CollectionOnline/SpecialCollectionRootFolder/32.

1.1.2 Detail Ha Bik-chuen studio

Courtesy of the Ha Family and Asia Art Archive
Photo: Jack Chueng of NOTRICH MEDIA, Hong Kong

1.1.3 Spread from one of Ha Bik-chuen's collage books

Courtesy of the Ha Family and Asia Art Archive

came to the territory, following the civil war in mainland China and had to survive in a place that was spatially not yet prepared for the influx of people.[87] As Hong Kong witnessed progress following its move to mechanical reproduction, so did Ha. Where Hong Kong's development was a direct result of the continuous production of things, which came with the kind of modern experience of boredom and the ever-present longing for the new, Ha's tinkering with the 'discarded' and the 'old' constructed an alternative experience of (or parallel to) a modern Hong Kong life.

His work and collages (and finally his archive) can be understood to prompt a different kind of sense experience in the same modern spaces of Hong Kong – sense experience not based on monotony but on a different dealing with time. Ha's sourcing of images, capturing of art shows, and subsequent putting together of collages, involve an oeuvre, not a dismissal of the 'old' or perpetual desire for the new. Yet, as his collection expanded in his library/studio at the rooftop of the Chinese tenement building in Kowloon East where he lived,[88] it also symbolises coexistence with the

87 Steve Tsang, *A Modern History of Hong Kong: 1841-1997* (London: L.B. Tauris & Co Ltd, 2007), 140.

88 Bernice Chan, 'A New Found World'.

rhythms of the modern city, including the very particular spatiality that modern Hong Kong developed as it defined its global position. Bourriaud argued,

> Art tends to give shape and weight to the most invisible processes. When entire sections of our existence spiral into abstraction as a result of economic globalization, when the basic functions of our daily lives are slowly transformed into products of consumption [...], it seems highly logical that artists might seek to rematerialize these functions and processes, to give shape to what is disappearing before our eyes. Not as objects, which would be to fall into the trap of reification, but as mediums of experience: by striving to shatter the logic of the spectacle, art restores the world to us as an experience to be lived.[89]

With Hong Kong's spectacle as a backdrop, the collection of Ha Bik-chuen proposes an experience not just of art, but also of living while his practice as archive suggests passion beyond boredom.

Life is Art: Frog King

Avant-garde artist Kwok Mang-ho (Frog King)'s studio is described by one of the curators of his 2011 Venice Biennale exhibition as a kind of 'archive-exhibit' that, using the clutter of randomly found objects, represents Hong Kong's 'high-density living environment'.[90] When people visit him, he tells them his story in the spatial context of his studio that 'forms a paradoxical landscape in which he uses found objects [...] to mock as well as transcend the materialistic world'.[91] Like Ha Bik-chuen, Frog King is a collector of sorts yet where Ha's archive presents a certain mode of living that comes with constant assembling and organising, Frog King pursues his motto, 'art is life, life is art'.

Guangdong-born and Hong Kong-raised, Frog King (1947) has been making art for nearly five decades, fifteen years of which he spent in the United States. Together with a number of other Chinese artists (including Ai Weiwei), Frog King was trained in New York, in the 1980s, before which he

89 Bourriaud, *Postproduction*, 32.
90 Tak-ping Tsang, '"Art is Life, Life is Art": Frog King's Art and Life as a Secular Pilgrimage of Illumination', in *La Biennale di Venezia Catalogue: Frogtopia Hongkornucopia* (2011), 22.
91 Ibid.

had received traditional arts training in Hong Kong.[92] Indeed, Frog King is of the generation that benefitted from Hong Kong's industrial development, which indirectly contributed to his exposure to the Western art scene. As he picked up 'the avant-garde spirit of Western modernist art',[93] he started to mix his traditional calligraphy with whatever materials he could find around him. This led to highly conceptual work with a distinct graphic style.

To Frog King, 'creation is a process of endless creative loitering, whether it happens to involve the use of traditional techniques, new technology, high-end or popular culture, conventional or unconventional practices, contemporary elements, or the detritus of everyday life'.[94] His exhibitions and performances exemplify this kind of mentality as his motto, 'art is life, life is art', is best taken as an 'experience'. It is how Rancière understands 'aesthetic experience', as it communicates 'the realm of art with that of life experience'.[95] Frog King's installations are 'spatial collages' that are made out of anything he has come across in the spaces of the city or in those of his studio: a redistribution of the sensible.

While his studio space and the things that can be found in it form a big segment of his 'art is life, life is art', his exhibition at the 2011 Venice Biennale titled *Frogtopia·Hongkornucopia* expounds his ideas in one show. One of the four sections that his exhibition covered was *Nine million works+'*: a space sprawled with a mix of materials and found objects, brought over from Hong Kong and reworked into his distinct aesthetic ensuing from his calligraphy background, featuring black, white, and red.

His exhibition involved a combination of installation and performance art. Indeed, Frog King often appears in his signature costume, acting out improvised performances that frequently also involve his audience for that his 'art is life' is about play. Customised glasses that he hands out allow his audiences to partake in his work and 'see' through his eyes. As the title implies, Frog King addressed the cluttered reality of urban life: 'Hongkornucopia' hints at some form of growing forest (a spatial disease) reflecting life in an overcrowded place – 'cornucopia' referencing an abundant supply of things. 'Frogtopia' is a 'counter-place' and 'counterculture' where play is

92 ArtAsiaPacific, 'Frog King Kwok Mang-Ho', *ArtAsiaPacific Magazine* 93 (2015), accessed December 4, 2016, http://artasiapacific.com/Magazine/93/FrogKing; Mang-ho Kwok, 'Artistic Concept and Artist Biography', *La Biennale di Venezia Catalogue: Frogtopia Hongkornucopia* (2011): 6.

93 Tak-ping Tsang, '"Art is Life, Life is Art"', 20.

94 Ibid., 21.

95 Ruben Yepes, 'Aesthetics, Politics, and Art's Autonomy: A Critical Reading of Jacques Rancière', *Evental Aesthetics* 3.1 (2014): 45.

key.[96] The latter can be seen as a place existing in a utopic future; a place that feeds off – or responds to – that what necessitates it: the modern city of Hong Kong and the monotony that came with its capitalist mode of production.

This responds to Rancière's idea of the aesthetic regime of art. Frog King's pursuit of 'art is life, life is art' helps examine boundaries between 'what is supposed to be normal and what is supposed to be subversive, between what is supposed to be active, and therefore political, and what is supposed to be passive or distant, and therefore apolitical'.[97] Like his studio, *Frogtopia·Hongkornucopia* is a place 'where one circulates differently between things, images, and words; there are tempos, a slow pace, a pause; there are arrangements of signs, a bringing together of distant things, schisms with united things'.[98] It does not critique consumer culture or the 'spatial disease'. It reorganises its elements and proposes a place for alternative, sensible experience beyond monotony.[99]

Like Ha Bik-chuen's studio, Frog King's may be perceived as an archive of life and art – of his shows and performances, of his life, and of other people's lives as projected in those things that the space encapsulates. I would have taken it as an archive of 'things' from the past, yet, the pursuit of 'life is art' does not fit such an idea and makes it a living 'milieu': a 'topos' within which both art and life happen.[100] The point is, however, that in their collection and assemblage, both artists – one of a generation that had to struggle through Hong Kong's post-war reality and the other profiting from the successes the previous generation had made – have gone beyond the ordering system of modernity. Both their work seems to be based on the premise that there is no division between art and life, nor between art and clutter. When 'Art is Life, Life is Art', boredom is reinvented.

Boredom and the Future: Aesthetics of Overabundance

Where in the everyday of a modern city such as Hong Kong, clutter and boredom have been continuously negotiated for decades, the two artists' oeuvres both suggest that there is an aesthetic alternative that – when

96 Benny Chia, 'Frogtopia-Hongkornucopia: Reading Frog King Kwok', *La Biennale di Venezia Catalogue: Frogtopia Hongkornucopia* (2011): 10.
97 Rancière, 'Art of the Possible', 266.
98 Ibid., 264.
99 Anneke Coppoolse, *Trash, a Curatorial: An Ethnographic and Visual Study of Waste in Urban Hong Kong*. Unpublished PhD dissertation (Hong Kong: Lingnan University, 2015), 254.
100 Ibid., 262.

keeping in mind Frog King's utopic interpretation of the spatial disease he encounters while living in Hong Kong – could be called an aesthetics of overabundance. Ha Bik-chuen's archival and collage approach to collection, further, similarly suggests a reinterpreting of vast amounts of things, in the case of his archive allowing a unique perspective on the history of the Hong Kong art scene since the 1960s.[101]

Both artists, however, should be understood in relation to the avant-garde tradition and have therefore, in Rancière's view,[102] imagined 'a life to come' in response to modern society's need for the eternal return of the same.[103] Chaotic display – either inspired by avant-garde work or presented in contemporary art where the flea market has become the dominant visual model[104] – still needs to find appreciation outside the spaces of the art word. 'The bored' of today, those who are still subject to the logic of commodity production, ought to find new ways of engaging with clutter towards a situation of play. Only then will proper aesthetic experience be possible and creativity return. In Hong Kong, where order and the division between commodity and waste has been instructed so significantly, clutter presents itself in variant locations and situations, yet its place between trash and society still needs to be acknowledged beyond the spaces of art. An embrace of boredom may happen once overabundance is experience proper, disorder is an accepted (if not celebrated) aspect of modern living, and the classifying broom turns into just another object.

Bibliography

Allon, Fiona. 'Green Renovations, Eco-Homes and Sustainable Home Improvement'. In *Ethical Consumption: A Critical Introduction,* eds. Tania Lewis and Emily Potter (London; New York: Routledge, 2010), 202-15.

Ammons, Archie Randolph. 'Garbage'. In *Best of the Best American Poetry: 25th Anniversary Edition*, eds. David Lehman and Robert Pinsky (New York: Scribner Poetry, 2013): 3-22.

ArtAsiaPacific. 'Frog King Kwok Mang-Ho', *ArtAsiaPacific Magazine* 93 (2015). Accessed December 4, 2016. http://artasiapacific.com/Magazine/93/FrogKing.

101 Bernice Chan, 'A New-Found World'.

102 Rancière, *Aisthesis*, 22-23.

103 Salzani, 'The Atrophy of Experience', 134.

104 Bourriaud, *Postproduction*, 28.

ArtPace. 'Minimal Factory ($1 Market)/Red Bull Party (with D.J.)', *Spring 2002 International Artist-in-Residence Program*. Accessed September 30, 2016. http://www.artpace.org/artists_and_curators/surasi-kusolwong.

Asia Art Archive. 'Ha Bik Chuen Archive'. Accessed September 24, 2016. http://www.aaa.org.hk/Collection/CollectionOnline/SpecialCollectionRootFolder/32.

Bauman, Zygmunt. *Modernity and Ambivalence* (Cambridge: Polity Press, 1991).

Benedict, Carol. *Bubonic Plague in Nineteenth-Century China* (Stanford, CA: Stanford University Press, 1996).

Benjamin, Walter. *The Writer of Modern Life: Essays on Baudelaire* (Cambridge: The Belknap Press, 2006).

—. 'Unpacking My Library: A Talk about Book Collecting'. In *Illuminations*, ed. Hannah Arendt (New York: Schocken Books, 1968), 59-68.

Boscagli, Maurizia. *Stuff Theory: Everyday Objects, Radical Materialism* (New York; London: Bloomsbury, 2014).

Bourriaud, Nicolas. *Postproduction*. 2nd edition (New York: Ram Publications & Dist, 2005 [2002]).

Bulow, Jeremy. 'An Economic Theory of Planned Obsolescence', *The Quarterly Journal of Economics* 101.4 (1986): 729-49.

Chakrabarty, Dipesh. *Habitations of Modernity: Essays in the Wake of Subaltern Studies* (Chicago, IL: University of Chicago Press, 2002).

Chan, Bernice. 'A New-Found World: Hong Kong Artist Ha Bik-chuen's Legacy', *South China Morning Post*, June 9, 2015. Accessed September 24, 2016. http://www.scmp.com/lifestyle/arts-entertainment/article/1817056/new-found-world-hong-kong-artist-ha-bik-chuens-legacy.

Chan, Man-hung. 'Ha Bik-chuen – Sentient Beings: Solo Exhibitions of Relief Prints'. In *Catalogue of Solo Exhibition* (Hong Kong: China Oil Painting Gallery Ltd., 1997), 6-7.

The Colony of Hong Kong. *Ordinance No. 5: Good Order and Cleanliness* (1844). Accessed August 23, 2016. http://oelawhk.lib.hku.hk/archive/files/2b4c38ded8e441159b4dcb8f05f1073d.pdf.

Chia, Benny. 'Frogtopia-Hongkornucopia: Reading Frog King Kwok', *La Biennale di Venezia Catalogue: Frogtopia Hongkornucopia* (2011): 10-5.

Clapp, Jennifer. 'The distancing of waste: Overconsumption in a global economy'. In *Confronting Consumption*, eds. Thomas Princen, Michael Maniates, and Ken Conca (Cambridge, MA: The MIT Press, 2002), 155-76.

Coppoolse, Anneke. *Trash, a Curatorial: An Ethnographic and Visual Study of Waste in Urban Hong Kong*. Unpublished PhD dissertation (Hong Kong: Lingnan University, 2015).

de Certeau, Michel, Luce Giard, and Pierre Mayol. *The Practice of Everyday Life, Vol. 2: Living and Cooking* (Minneapolis, MN: University of Minnesota Press, 1998).

de Certeau, Michel. *The Practice of Everyday Life* (Berkeley, CA: University of California Press, 1984).

Dezeuze, Anna. 'Assemblage, Bricolage, and the Practice of Everyday Life', *Art Journal* 67.1 (2008): 31-37.

Donald, David. *A Financial Centre for Two Empires: Hong Kong's Corporate, Securities and Tax Laws in its Transition from Britain to China* (Cambridge: Cambridge University Press, 2014).

Drainage Services Department. *Sewerage and Flood Protection: Drainage Services 1841-2008* (Hong Kong: Drainage Services Department, The Government of Hong Kong Special Administrative Region, 2008). Accessed July 15, 2016. https://www.dsd.gov.hk/EN/Files/publications_publicity/Publications/2008book.pdf.

Environment Bureau. *Blueprint for Sustainable Use of Resources 2013-2022*. (Hong Kong: Environment Bureau, The Government of Hong Kong Special Administrative Region, 2013). Accessed July 15, 2016. http://www.enb.gov.hk/en/files/WastePlan-E.pdf.

Gardiner, Michael. 'Henri Lefebvre and the "Sociology of Boredom"', *Theory, Culture & Society* 29.2 (2012): 37-62.

Grossberg, Lawrence. *Cultural Studies in the Future Tense* (Durham, NC: Duke University Press, 2010).

Ha, Bik-chuen. *Explorer – The Art of Ha Bik-chuen* (Hong Kong: Artrend, 2004).

—. *Ha Bik-chuen: Ink at 76 (Exhibition and Sale)* (Hong Kong: Grotto Fine Art Ltd, 2001).

Hawkins, Gay. 'Waste in Sydney: Unwelcome Returns', *PMLA* 122.1 (Special Topic: Cities) (2007): 3348-351.

—. *The Ethics of Waste: How We Relate To Rubbish* (Lanham, MD: Rowman and Littlefield Publishers, 2006).

Hetherington, Kevin. 'Secondhandedness: Consumption, disposal, and absent presence', *Environment and Planning D – Society and Space* 22 (2004): 157-73.

Highmore, Ben. *Ordinary Lives: Studies in the Everyday* (London; New York: Routledge, 2011).

HKSAR. 'Amendment Regulation to Regulate Export of Powdered Formula Gazetted', *Press Releases*, February 22, 2013. Accessed July 20, 2016. http://www.fhb.gov.hk/gia/general/201302/22/P201302220597.htm.

Hoffman, Eva. *How to Be Bored* (London: Macmillan, 2016).

Hong Kong Memory. 'Industrialization in Post-war Hong Kong', *Hong Kong Memory* (website), 2012. Accessed January 4, 2017. http://www.hkmemory.hk/collections/postwar_industries/industrialization_in_postwar_hong_kong/index.html#&panel1-1.

Information Services Department. 'Of Green Monsters and Broken Daggers', *GIS Through the Years*, 1999. Accessed August 2, 2016. http://www.info.gov.hk/isd/40th/13.html.

Jennings, Michael. 'Introduction'. In *The Writer of Modern Life: Essays on Baudelaire*, ed. Walter Benjamin (Cambridge: The Belknap Press, 2006), 1-25.

Kwok, Mang-ho. 'Artistic Concept and Artist Biography', *La Biennale di Venezia Catalogue: Frogtopia Hongkornucopia* (2011): 6-8.

Lam, Anne. *Explorer – The Art of Ha Bik-chuen* (Hong Kong: Artrend, 2004).

Lehman, David. 'A.R. Ammons, The Art of Poetry No. 73', *The Paris Review* 139 (1996). Accessed December 22, 2016. https://www.theparisreview.org/interviews/1394/a-r-ammons-the-art-of-poetry-no-73-a-r-ammons.

Lee, Eliza, Elaine Chan, Joseph Chan, Peter Cheung, Wai-fung Lam, and Wai-man Lam. *Public Policymaking in Hong Kong: Civic Engagement and State- Society Relations in a Semi-Democracy* (Oxford: Routledge, 2013).

Lefebvre, Henri. *Critique of Everyday Life* (London: Verso, 2014).

—. *Rhythmanalysis: Space, Time, and Everyday Life* (London: Continuum, 2004).

—. *Critique of Everyday Life. Vol. II* (London: Verso, 2002).

—. *Everyday Life in the Modern World* (London: Continuum, 2002).

—. *Writings on Cities* (Oxford: Basil Blackwell, 1996).

—. *Introduction to Modernity: Twelve Preludes, September 1959-May 1961* (London: Verso, 1995).

Laidler, Karen Joe, and Maggy Lee. 'Border Trading and Policing of Everyday Life in Hong Kong'. In *The Routledge Handbook on Crime and International Migration*, eds. Sharon Pickering and Julie Ham (London and New York: Routledge, 2015), 316-28.

McCarthy, Anna. 'Clutter and Clodder', *Social Text* 34.3 (128) (2016): 127-29.

Moran, Joe. 'Benjamin and Boredom', *Critical Quarterly* 45.1-2 (2003): 168-81.

NYC Garbage. 'The Story behind HYC Garbage'. Accessed October 3, 2016. http://nycgarbage.com/about/.

OED Online. s.v. 'Waste'. Accessed September 25, 2016. https://en.oxforddictionaries.com/definition/waste.

OED Online. s.v. 'Clutter'. Accessed September 25, 2016. https://en.oxforddictionaries.com/definition/waste.

Papastergiadis, Nikos. *Spatial Aesthetics: Art, Place, and the Everyday* (Amsterdam: Institute of Network Cultures, 2010).

Peckham, Robert. 'Hong Kong Junk: Plague and the Economy of Chinese Things', *Bulletin of the History of Medicine* 90.1 (2016): 32-60.

Pezze, Barbara and Carlo Salzani. 'Introduction'. In *Essays on Boredom and Modernity*, eds. Barbara Pezze and Carlo Salzani (Amsterdam: Rodopi, 2009): 5-33.

'The Plague in Hong Kong', *Times*, August 28, 1894, p.6.

Princen, Thomas, Michael Maniates, and Ken Conca, eds. *Confronting Consumption* (Cambridge, MA: MIT University Press, 2002).

Rancière, Jacques. *Aisthesis: Scenes from the Aesthetic Regime of Art* (London: Verso, 2013).

—. 'Art of the Possible: Fulvia Carnevale and John Kelsey in Conversation with Jacques Rancière', *Artforum* 45.7 (2007): 256-69.
—. *The Politics of Aesthetics: The Distribution of the Sensible* (London: Bloomsbury Academic, 2004).
Salzani, Carlo. 'The Atrophy of Experience: Walter Benjamin and Boredom'. In *Essays on Boredom and Modernity*, eds. Barbara Pezze and Carlo Salzani (Amsterdam: Rodopi, 2009): 127-54.
Schaffner, Ingrid. 'Jason Rhoades, Four Roads: A Case Study of Contemporary Art and Conservation', *Voices of Contemporary Art Journal* (2015). Accessed January 8, 2017. http://journal.voca.network/jason-rhoades-four-roads/.
Smith, Dennis. *Zygmunt Bauman: Prophet of Postmodernity* (Cambridge: Polity Press, 1999).
The Museum of Modern Art. 'Surasi Kusolwong'. 2015. Accessed October 4, 2016. http://momaps1.org/exhibitions/view/345.
—. *Press Release: The Art of Assemblage* (New York: MoMA, 1961).
Tsang, Steve. *A Modern History of Hong Kong: 1841-1997* (London: L.B. Tauris & Co Ltd, 2007).
Tsang, Tak-ping. '"Art is Life, Life is Art": Frog King's Art and Life as a Secular Pilgrimage of Illumination'. In *La Biennale di Venezia Catalogue: Frogtopia Hongkornucopia* (2011), 20-9.
Vigneron, Frank. *Understanding Hong Kong Art through Hong Kong Culture.* Art Appreciation and Criticism in Context Series (Hong Kong: The Government of Hong Kong Special Administrative Region, Education Bureau, 2012).
Whiteley, Gillian. *Junk: Art and the Politics of Trash* (London: I. B. Tauris, 2010).
Yepes, Ruben. 'Aesthetics, Politics, and Art's Autonomy: A Critical Reading of Jacques Rancière', *Eventual Aesthetics* 3.1 (2014): 40-64.
Yip, Ka-che. 'Segregation, Isolation, and Quarantine: Protecting Hong Kong from Diseases in the Pre-war Period', *Journal of Comparative Asian Development* 11.1 (2012): 93-116.

About the Author

Anneke Coppoolse is an Assistant Professor in the School of Design at the Hong Kong Polytechnic University. She focuses on urban visual culture and the visuality of abandoned objects, navigating questions of aesthetics and lived experience.

1.2 Combating the Boredom of Traditional Performing Arts?

The Case of *Muyuge*

Wen Cuiyan

Abstract

In the fever surrounding Intangible Cultural Heritage (ICH), there is much debate about whether traditional performing art forms should be more creative in order to remain sustainable in modern times. This paper interrogates the relations between China's ICH policy, local practitioners, and the 'creativity' of traditional performing art forms. *Muyuge*, a genre of narrative singing, used to be popular in Cantonese-speaking regions. In 2011, it was inscribed as a national ICH project. However, the actual everyday practices remain a neglected realm. Based on comparative analyses of *muyuge* practices by the official ICH sectors and ordinary people, Cuiyan argues that 1) the current 'creative' practices marginalise local practitioners; 2) the 'creative' traditional performing forms cannot last when detached from everyday contexts.

Keywords: Intangible Cultural Heritage, boredom, Cantonese, performing arts, *muyuge*

Huang Peiyi, born in the late 1980s in Guangdong, a southern province of China, still vividly remembers her first encounter with *muyuge* (木鱼歌).[1] It was in 2002, when she was a high school student. In one of the regular class meetings, her head teacher, Li Zhongqiu, excitedly shared a song he wrote for the class. Mr. Li showed the lyrics on the projector. The class instantly

1 In this essay, I use the Putonghua-based pinyin, the official Romanisation system for Standard Chinese in mainland China.

Kloet, Jeroen de, Chow Yiu Fai, and Lena Scheen (eds), *Boredom,* Shanzhai, *and Digitisation in the Time of Creative China*. Taylor & Francis Group, 2019
DOI: 10.5117/9789462984745_CH1.2

erupted as the students saw that the lyrics were actually composed of the 52 names of the class. The class descended further into chaos when Mr. Li explained he intended to teach them to sing it. 'We all thought the song was boring and weird. The pace was so slow and the tune so strange', Huang recollects. 'Growing up with music like Korean pop and Western pop, we thought this kind of song was simply unacceptable'. The song was never sung again. What Huang and her classmates did not know was that her teacher, once an enthusiastic fan of *muyuge*, wrote the song to the tune of *muyuge*, a genre of Cantonese narrative singing that had been popular in the Cantonese-speaking area from the late Ming period (1368-1644).

If it were not for a high-profile campaign relating to intangible cultural heritage (ICH), *muyuge* may have disappeared without a trace. After China ratified the Convention for the Safeguarding of the Intangible Cultural Heritage in 2004 and set up its own national system of protecting intangible cultural heritage, a total of 1,519 cultural practices across the country were recognised as national intangible heritage.[2] Moreover, numerous ICH projects are generated at the provincial, municipal, and county levels every year. The town of Dongkeng in Dongguan City 're-discovered' *muyuge* as an intangible cultural heritage in 2006 and proceeded to inscribe it in the national inventory in 2011. Despite the national title, local ICH officials still struggle to re-root this cultural practice in the very place where it used to prosper. The contradictory moment of the teacher's enthusiasm and his students' rejection reveals a difficult gap in the aesthetic tastes of different generations. In response to the resurgence of *muyuge* as intangible cultural heritage, not only do the authorities call for the preservation of a traditional form, but also a modern representation and transformation.

2 At time of writing, (January, 2017), the State Council of the People's Republic of China has announced four lists of national intangible cultural heritage successively in 2006, 2008, 2011, and 2014: 'The State Council's Notice about the Fourth Group of National Intangible Cultural Heritage Inventory [国务院关于公布第四批国家级非 物质文化遗产代表性项目名录的通知]', November 11, 2014, accessed January 23, 2017, http://www.gov.cn/zhengce/content/2014-12/03/content_9286.htm; 'The State Council's Notice about the Third Group of National Intangible Cultural Heritage Inventory [国务院关于公布第三批国家级非物质文化遗产代表性 项目名录的通知]', May 23, 2011, accessed January 23, 2017, http://www.gov.cn/zwgk/2011-06/09/content_1880635.htm; 'The State Council's Notice about the Second Group of National Intangible Cul tural Heritage Inventory and the Expanded Items of the First Group of National Intangible Cultural Heritage Inventory [国务院关于公布第二批国家级非物质 文化遗产名录和第一批国家级非物质文化遗产扩展项目名录的通知]', June 7, 2008, accessed January 23, 2017, http://www.gov.cn/zwgk/2008-06/14/content_1016331.htm; 'The State Council's Notice about the First Group of National Intangible Cultural Heritage Inventory [国务院关于公布第一批国家级非物质文化遗产名录的通知]', May 20, 2006, accessed January 23, 2017, http://www.gov.cn/zhengce/content/2008-03/28/content_5917.htm.

Muyuge is not a unique case. Since the beginning of the new millennium, many indigenous, everyday traditions have been highlighted by the new concept of intangible cultural heritage around the globe. Increasing interest in traditional practices has been stirred up. On the other hand, this heated interest is always accompanied by controversy and dissatisfaction due to the 'boring' experience of traditional practices, especially when compared with modern popular culture. What does the bored feeling of traditional cultural practices signify? Why does the government believe it is so important to reverse any feelings of boredom? In this chapter, I take *muyuge* as a case study to examine these questions. I have conducted several interviews with Chinese ICH officials as well as participant observations of local events and courses concerning *muyuge* in 2015 and 2016. In this chapter, I examine the public sector's effort to tackle 'boringness' under the framework of intangible cultural heritage in contemporary China.

Intangible Cultural Heritage

Before going into the details of the case study, I would like to trace the history of the emergence of heritage, especially intangible cultural heritage, and seek to position the feeling of boringness towards traditional cultural practices within this context. The idea of heritage emerged in Europe and the U.S. during the nineteenth century and it is closely associated with nation-building and rapid industrial development. In the twentieth century, a Western-centric discourse of was gradually constructed, which Laurajane Smith (2006, p. 11) coined the 'authorised heritage discourse' (AHD).[3] The Convention Concerning the Protection of the World Cultural and Natural Heritage adopted by UNESCO in 1972 institutionalised the discourse by setting up an international heritage agenda based on Western values and thought systems. Heritages recognised by this discourse usually convey a sense of monumentality, pleasing aesthetics, and are inherently associated with human nature. Thus, AHD privileges the material reality of heritage as well as voices of expertise, while broader public engagement is limited.[4] The canonical perception of heritage was disputed due to the emergence of 'unofficial heritages' in Europe and North America and critiques of the cultural dominance of the West from Third World countries and ethnic groups in

3 Laurajane Smith, *Uses of Heritage* (New York: Routledge, 2006), 11.

4 Smith, *Uses of Heritage*; 'Intangible Heritage: A Challenge to the Authorised Heritage Discourse?', *Revista d'etnologia de Catalunya* (2015), 133-42.

the 1980s. 'Unofficial heritage' refers to practices that are considered as heritage but not officially recognised.[5] Due to the prevalence of cultural tourism under the trend for neo-liberal economy, a vast array of popular culture and everyday practices have been absorbed into heritage in order to attract a mass audience.[6] Also in the 1980s and 1990s, increasing ethnic groups gained independence. There was a greater demand for 'the expression of local identity through traditional cultures'. Meanwhile, intellectual criticism of the West-centric understanding of heritage emerged and became stronger.[7] The latter led to the implementation of the Convention for the Safeguarding of the Intangible Cultural Heritage in 2003.

The 2003 Convention broadened the scope of heritage to the intangible domain, i.e. oral traditions, performing arts, and other everyday practices. It also attempted to raise awareness of local and indigenous cultures as well as to empower relative communities and sub-nation groups. There were high expectations that the Convention would challenge the authorised heritage discourse. However, it has been increasingly criticised as ICH has failed to meet such expectations. Like World Cultural Heritage, ICH also operates under UNESCO and is based on state members, which enables the state to appropriate local heritages as state resources. Furthermore, professionals continue to play a crucial role in evaluating and guarding the thresholds of the ICH lists. The management system poses a real challenge to the authorised discourse of heritage.[8] ICH follows the same logic that has shaped the concept of world heritage. In other words, AHD is still shaping the way people perceive and experience intangible heritage. In fact, the majority of intangible cultural heritage derives from everyday cultural practices that ordinary people and marginal communities (used to) practice but which were unrecognised by mainstream society. The ordinariness of these practices makes it hard to meet the heritage standard set by the AHD. These frictions strengthen the heritage perception and experience defined by the AHD but make it harder to accept the ordinariness and indigenous aesthetics that fall outside of the definition of AHD. I argue that it is this unwillingness and difficulty to accept traditional practices in a contemporary context that leads to feelings of boredom and estrangement

5 Rodney Harrison, *Heritage: Critical Approaches* (London; New York: Routledge, 2013), 15.

6 Harrison, *Heritage*; Bella Dicks, *Culture on Display, the Production of Contemporary Visitability* (Maidenhead: Open University Press, 2004).

7 Natsuko Akagawa, 'Intangible Heritage and Embodiment: Japan's Influence on Global Heritage Discourse'. In *A Companion to Heritage Studies,* eds. William Logan, Mairead Nic Craith, and Ullrich Kockel (Chichester: Wiley Blackwell, 2015), 69-86.

8 Smith, 'Intangible heritage'.

towards older cultural practices. In order to engage traditional cultural practices in the contemporary cultural scape, the authorities must make give such practices a 'less boring', more 'modern' narrative.

China has been relatively late to engage with heritage preservation and tourism. In the 1960s, traditional architecture and many cultural practices were considered 'the Four Olds' and were to be eliminated.[9] It was not until 1985 that the Chinese government officially ratified the World Heritage Convention. During the 1980s, culture, not to mention heritage, was still considered as anti-market and dominated by social elites. In the 1990s, the Party-state began to promote leisure culture as a kind of popular lifestyle and a means to stimulate domestic demands. The nationwide implementation of the double leisure day system[10] in 1995 and three 'golden weeks'[11] in 1999 marked an upsurge in tourism. Heritage sites became hot touristic destinations. When the idea of intangible cultural heritage emerged in the early 2000s, China quickly picked it up. In accordance with the Convention for the Safeguarding of the Intangible Cultural Heritage 2003, the Chinese government drew up a domestic ICH inventory system that covered national, provincial, municipal and county levels.

The ICH campaign is a combination of the political and economic agendas of the Party-state. Politically, it is a continuation of the Party-state's political strategy since 1989 of shifting the ideological emphasis from communism to patriotism and national traditions. In the 1990s, the leisure culture campaign was not only promoted for economic purposes, but also incorporated into the political agenda of educating residents to become 'modern, cultured, and public citizens' in order to construct 'socialist spiritual civilization'.[12] Over the past twenty years, socialist spiritual civilisation has morphed into a cultural nationalistic narrative. In 2013, Chinese President Xi Jinping proposed the national ideals of 'China's dream' and 'the rejuvenation of the Chinese nation', which places significant focus on traditional cultures. A State Council document about reinforcing the preservation of intangible

9 'The Four Olds' stands for old customs, old culture, old habits, and old ideas.

10 In Chinese '双休日'。China began to implement the five-day working week system on 1 May, 1995, establishing Saturday and Sunday as official rest days.

11 In Chinese '黄金周'。In 1999, the official holidays of Labour Day, National Day, and the Spring Festival were each extended to three days. Together with the weekends before and after the official holidays, people are able to enjoy three seven-day holidays in a year. People call these long holidays the 'golden weeks'. In 2008, the official Labour Day holiday was shortened to one day.

12 Jing Wang, 'Culture as Leisure and Culture as Capital', *Positions: East Asia Cultures Critique* 9.1 (2001): 69-104.

cultural heritage in 2005 reads, 'The intangible cultural heritage of our country embodies the unique spiritual values, ways of thinking, imagination and cultural awareness. It is the fundamental basis to maintain the cultural identity and cultural sovereignty of our country'.[13] In other words, the boom of ICH since the early 2000s is an active response to this nationalistic agenda.

Economically, the ICH campaign opens up vast space and potential for the exploitation of cultural tourism and a cultural economy. The years around 2004 witnessed the first crisis of migrant workers in the economically developed Pearl River Delta area, resulting in critical discussions calling for economic restructure and upgrading.[14] At the same time, possibilities that might lead to a greener and more efficient economic model were conceived. In the early 2000s, the significance of a knowledge-based economy was widely acknowledged in society. The term 'cultural industries' was formally adopted in China's Tenth Five-Year Plan (2001-2005). In the meantime, there was growing attention to intangible heritage. Following the implementation of the UNESCO Convention for the Safeguarding of Intangible Cultural Heritage in 2003, China was quick to respond and ratify the Convention in 2004.[15] In 2006, the State Council announced the first national list of intangible cultural heritage. Over the past ten years, the Ministry of Culture has announced four national lists. Moreover, numerous ICH projects are generated every year at provincial, municipal, and county levels. Lower-level governments and certain enterprises, especially those from more economically developed regions, are most enthusiastic about engaging with ICH. Intangible heritage, alongside other cultural heritage, has become an efficient way to promote the cultural values of a city or an enterprise, and a potential boost for the local economy. In sum, traditional cultural practices have been rendered a different role in contemporary China as such practices represent the rejuvenation of the Chinese nation as well as a new boost for economic growth. Both the nationalistic narrative and the economic demand

13 The State Council of the People's Republic of China, 'The State Council's Notice about the First Group of National Intangible Cultural Heritage Inventory [国务院关于公布第一批国家级非物质文化遗产名录的通知]', May 20, 2006, accessed January 23, 2017, http://www.gov.cn/zhengce/content/2008-03/28/content_5917.htm.

14 Dunfeng Dai, 'China faces the first factory worker crisis during the past 20 years [中国遭遇 20 年来首次民工荒]', *Southern Weekly*, July 17, 2004, accessed October 3, 2016, http://www.aisixiang.com/data/3600.html; Kailai Xiang, 'From the wave of migrant workers to the crisis of migrant workers, what does it indicate?' ["民工潮"到"民工荒"，说明了什么?]', *Xinhua News*, September 6, 2004, accessed October 3, 2016, http://news.xinhuanet.com/focus/2004-09/06/content_1938673.htm.

15 UNESCO, 'Text of the Convention for the Safeguarding of the Intangible Cultural Heritage', 2003, accessed October 3, 2016, http://www.unesco.org/culture/ich/en/convention.

behind heritage development in China call for a spectacular representation of traditional performing arts and cultural practices, which follows the authorised discourse of heritage rather than the opposite. Focusing on the study of *muyuge*, the rest of this chapter provides more details of the tension between traditional cultural practices and the reconstruction of them as intangible heritage, as well as the tension between feelings of boredom and official attempts to combat such feelings in the contemporary social and cultural context of China.

Muyuge

Generally speaking, *muyuge* is a kind of narrative singing. Narrative singing is also called *quyi* (曲艺) and *shuochang* (说唱) in Chinese, which means to tell stories by both singing and speaking. This kind of performance has been popular across China for nearly a thousand years and there are 260 different regional variations.[16] *Muyuge* used to be popular in the Cantonese-speaking areas in Guangdong Province. It is generally composed in seven-syllable verses with a loose tonal and rhythmic pattern. Verses are performed in a half speech-like and half song-like manner. People sing *muyuge* for entertainment, religious events and rituals like wedding ceremonies, funerals. *Muyuge* is also considered as a kind of popular literature. As the publishing industry advanced around the late Ming (1368-1644) and early Qing (1644-1912), booklets of *muyuge* lyrics, also called *muyushu* (木鱼书, 'wooden-fish book'), emerged and were sold at low prices.[17] Due to high accessibility, booklets were widely circulated among ordinary people. Reading and singing *muyuge* booklets became a popular form of entertainment. Apart from the large number of ordinary practitioners, singing *muyuge* used to be a major profession for blind people until the mid-1980s. To increase performative effects, blind singers performed with *sanxian*, a three-string Chinese instrument that is plucked. Compared with most modern popular music, *muyuge* performances are much longer. The comparatively shorter pieces, also called *zejin* ('excerpts'), generally last for thirty minutes to an hour. Longer ones are performed for hours and even days. Moreover, it is sung in a speech-like

16 Bell Yung, *Cantonese Opera Performance as Creative Process* (Cambridge: Cambridge University Press, 1989), 138.

17 Pui-chee Leung, *Wooden-Fish Books: Critical Essays and an Annotated Catalogue Based on the Collections in The University of Hong Kong* [香港大学所藏木鱼书许录研究] (Hong Kong: Centre of Asian Studies, The University of Hong Kong, 1978), 206.

manner with very simple and repetitive rhythms and inflections. The classic style of singing that appreciates the inflections of tones and the nuances of emotional expressions is performed at much slower pace.[18] Before the flood of popular culture, *muyuge* was among the very few entertainments that ordinary people could access in peripheral towns and cities. Many people, like the high school teacher Li Zhongqiu mentioned at the beginning of the essay, appreciated and enjoyed the experience of *muyuge*.

Constant warfare in the first half of the twentieth century, especially the anti-Japanese War from the 1930s to 1940s, destroyed the social environment that sustained the stable cultural and entertainment life of ordinary people. The publication of *muyushu* began to decline. Even so, *muyuge* singing remained popular in everyday life. Many of my interviewees, in their sixties and above, recollected scenes of singing and listening to *muyuge* with their family, friends, and neighbours when they were young. Until the mid-1950s, local cultural departments still invited blind singers for public performances on important holidays. In 1951, a top-down campaign of drama reform swept the country. It was aimed at re-ordering social life and included modifying the old drama repertoire and operational system, replacing them with communist ideologies. As a minor and declining practice, *muyuge* was not fundamentally influenced in terms of everyday practices; however, it was overshadowed by perceptions of feudal superstition and ideological backwardness against the dominance of socialist ideology. In addition to the reinforcement of joint state-private ownership in 1956, small business withered and, consequently, *muyuge* lost major patrons. As the Cultural Revolution began, many blind singers stopped performing and the practice went underground. Moreover, many people burned their *muyushu* due to huge political pressure. It was not until the late 1970s, when the Cultural Revolution ended, that *muyuge* practices were revitalised on a small scale among enthusiasts. In the three decades that followed, however, the overwhelming prevalence of modern popular music and culture has eclipsed the already vulnerable voices. As enthusiasts get older and pass away, *muyuge* is gradually disappearing from the everyday soundscape of the city.

Muyuge has a strong sense of everydayness. By 'everydayness', I refer to the non-spectacular quotidian life of ordinary people. Before it was enrolled in the national intangible heritage list, very few people recognised

18 Classical singing was popular among the old literati circle before the early 1950s. Prior to the 1950s, blind singers were able to perform in the classical style; however, classical singing has now almost disappeared. Yang Baolin, an 80-year-old scholar from Dongguan, is the only person who is able to perform classical *muyuge* singing today.

muyuge as a kind of 'art' or even 'culture'. There was not even an official expression to describe such practices. Most practitioners and enthusiasts use unofficial expressions such as *ge* ('songs'), *geben* ('song booklets'), or *manglaoge* ('the blind men's song'). *Muyuge*, as an official term, was only circulated among a small academic circle. It was after its transformation into an intangible heritage that the term began to gain more popularity. Surprisingly, *muyuge* was more popular in peripheral towns and cities in Guangdong, such as Dongguan, Foshan, Enping, and Taishan, rather than in bigger Cantonese-speaking cities like Guangzhou, Hong Kong, or Macau. Many *muyuge* practitioners and enthusiasts came from lower social classes. Blind singers simply took *muyuge* as a way of surviving. For ordinary people, it was nothing but an everyday entertainment or a minor practice that went along with other routines like religious events and ceremonies. Considering the social and historical contexts of *muyuge*, especially the long ignorance of it by the elite class in the imperial time and its ideological exclusion in the socialist period, it is not difficult to observe a deep-rooted disdain towards vernacular everyday cultures from dominant cultural values. Such bias is also responsible for the later unwillingness to accept *muyuge* and the accusations that it is 'boring'.

'Building a Town of Exquisiteness and Distinction'

The town government of Dongkeng is officially responsible for the safeguarding of *muyuge* as an intangible cultural heritage. Dongkeng, one of the 32 town districts of Dongguan, is a typical Pearl River Delta town that has undergone radical urbanisation over the past thirty years. When I visited Dongkeng in August 2016, I saw a huge banner on an overpass that read 'Upgrade the development quality of Dongkeng; building a town of exquisiteness and distinction'.[19] Under the huge banner, there were a large number of factory recruitment stalls seeking welders, sewing workers, and other assembly line workers. In the late 1970s, the once agriculture-based town began to introduce foreign investment and set up many labour-intensive factories. However, in the past decade, these small, low-end factories have been confronted with a constant shortage of factory workers and decreasing policy support. At the same time, the town is actively seeking industrial transformation. Taking advantage of being adjacent to a high-tech industry development zone, Dongkeng strives to attract larger high-tech enterprises

19 In Chinese, '提升东坑发展质量，建设精品特色小镇'.

and shake off low-end factories in order to reach 'higher development quality'.[20]

On the other hand, the town government is attempting to exploit the potential of cultural tourism. The emergence of intangible cultural heritage offers a big opportunity. At the time of writing, Dongkeng has five recognised ICH projects: *muyuge*, the Maishen Festival, a kind of local dried radish, a local snack of sweet rice ball and a local spiritual practice. All are local cultural practices, among which *muyuge* reaches the highest level of national recognition. It is noteworthy that Dongkeng was not well known for practices of *muyuge* until 2005 compared to other towns of Dongguan.[21] However, it was the first town to submit application and to express strong willingness to preserve *muyuge* as a heritage. *Muyuge* thus ended up being a cultural resource for Dongkeng. The Dongkeng government had an even greater ambition to restructure the town's cultural map. In 2014, the town government proposed the concept of 'three ancient cultures', namely, ancient trees, ancient architectures and ancient folk customs, to promote the 'traditional agricultural culture' of Dongkeng.[22] *Muyuge*, together with other local intangible heritages, is included in the category of 'ancient folk customs'. 'Three ancient cultures' shows a new cultural imaginary and a systematic way to re-organise cultural resources, which emphasises the inherent and 'exquisite' cultural characteristics of the town.

From Everyday Practice to Public Performance

As argued earlier, *muyuge* is an everyday practice of the ordinary people. Conflicts emerge when this practice is granted the status of cultural heritage and is imbued with the expectation of adding to the values of 'a town of exquisiteness and distinction'. Local ICH officials aim to combat the audience's feeling of boredom[23] and reverse its image as a lower-class, marginal

20 Dongkeng has built a number of new industrial parks to take in ancillary investment from Shenzhen and Songshanhu Lake Industry Park. It is reported that by the end of 2015 Dongkeng had successfully introduced five major investment projects of 2.7 billion RMB, approximately 460 million USD, including high-end electronic technology and biological technology (*Southern Daily*, 2016).

21 In terms of the cultural resources of *muyuge*, other towns, such as Guancheng, Liaobu, and Daojiao, could have been more competitive than Dongkeng. It appears, however, they were slow to realise *muyuge*'s potential value as heritage and missed the chance to register it as such.

22 In Chinese, '打造传统农耕文化'. *Suno769*, 'Three ancient cultures boost agricultural park and create new "selling points" of leisure tourism ["三古"文化助推农业园打造休闲旅游新"卖点"]', December 17, 2015, accessed October 3, 2016, http://dk.suno769.com/jinrikandian/jinriyaowen/59988.shtml?1450342345.

23 Zhongqiu Li, 'How to Transmit Muyuge [如何做好木鱼歌的传承]', in *Selected Works of the Transmission and Preservation of Muyuge* (Dongkeng, China: the Muyuge Transmission Base

practice. *Muyuge* is presented as an intangible cultural heritage in two different settings. One is informal performances at events such as exhibitions and public lectures. The ICH inheritor[24], performs *muyuge* pieces, sometimes with his students. The repertoire includes some classic pieces, but most are new composition. All the pieces are short and performed within ten minutes. The other setting in which *muyuge* is performed is more formal and includes e.g. gala shows and folk-art competitions. These performances are more carefully designed and produced and better embody local government's perception of cultural heritage and their efforts to reconstruct *muyuge* as a modern performance. I will therefore focus on this latter presentation from the perspective of three tactics employed by the local cultural sector to reconstruct *muyuge*, i.e. festivalisation, transforming it into a modern stage performance, and making language changes.

Festivalisation

Town officials were quick to realise that it was not wise to promote *muyuge* only for the sake of *muyuge*. In 2008, they put it on the stage at the Maishen Festival thereby establishing a new tradition that *muyuge* is performed at the festival every year. The Maishen Festival is a local folk festival that originated from the labour force market in the Qing Dynasty and later developed into a market day combining everyday farm products with ritual celebrations. Town folks resumed the festival after the Cultural Revolution in the 1980s. In 2002, the town government began to integrate the festival into its cultural development scheme.[25] Since then, the Maishen Festival has become the highlight of the year in Dongkeng. The festival[26] features galas, fairs, parades, and water carnivals where people across the town splash water at each other. It is an important event in terms of attracting investment and tourists. Intangible heritage is a key theme of the festival. Organisers

of Dongkeng Dongguan, 2009), 34-5; Pinrui Culture, 'Exploration and Transmission: Dongguan Muyuge [探寻传承之东莞木鱼歌]', 2014, accessed October 3, 2016, http://www.dgswhg.com/fy/sub/?id=135.

24 China established the nationwide ICH inheritor system in the wake of the ratification of the 2003 Convention. An ICH inheritor is officially selected and recognised for his/her skills, engagement and reputation of a certain ICH project. The inheritor receives governmental funding and is responsible for the transmission of the heritage project. The ICH inheritor system is similar to the former UNESCO programme of Living Human Treasure.

25 *Dongkeng Annals* (Guangzhou: Lingnan Fine Arts Publishing House, 2008).

26 The festival is on the second day of the second lunar month, usually at the end of February and March.

invite cultural products and performances from Dongkeng and other places for trade and exhibition. *Muyuge* is the highlight of the festival's opening ceremony and the town government invests a lot to invite professional composers, directors, and actors to put on new *muyuge* performances.

The opening ceremony takes place at Dongkeng Century Square, which is located on the opposite side to the town hall in the new town centre. The central stage is more than twenty metres wide. As the ceremony is open to thousands of visitors every year, the traditional form of *muyuge* is considered too simple for such a big stage. In accordance with the atmosphere of the Maishen Festival, new *muyuge* performances are specifically created in a festive and celebratory style. The stage bustles with people, sounds, and colours so that it catches the audience's eyes. New *muyuge* performances publicise the cultural attractions of Dongkeng. For example, the performance of the lyrics of *Maishen Festival sung in Muyuge* tell how the Maishen Festival has transformed from a sad story of labourers in the past to a festival for all Dongkeng people today; *Sweet sweet rice ball* sings of the local snack, another intangible heritage of Dongkeng; *Evening singing at the Temple of Tinggang* (abbr., *Evening singing*) is in praise of the landscape around the Temple of Tinggang, one of the government-promoted tourist attractions. The aim is to embed *muyuge* in a larger network of cultural resources so as to create bigger impact.[27]

Modern Stage Performance

The *muyuge* performed on stage is always a mix of dancing, singing, and even theatre. *Evening singing* is considered a piece that maintains the most 'authentic' *muyuge* elements. Li Zhongqiu, the *muyuge* inheritor (the high school teacher mentioned earlier), composed the lyrics in neat seven-syllable verses with orderly tonal and rhythmic patterns. In order to polish the musical form, the town government invited a popular composer[28] to integrate it with pop music, and later professional director, actors, and dancers to complete the whole production. At the opening ceremony of the 2009 Maishen Festival, the performance of *Evening singing* was led by the *muyuge* inheritor and accompanied by around twenty women performers.

27 Duzhong He, 'Transmitting Muyuge, Five Easy Points, Five Difficult Points and Two Suggestions [传承木鱼歌的五易五难和两点建议]', in *Selected Works of the Transmission and Preservation of Muyuge* (Dongkeng, China: The Muyuge Transmission Base of Dongkeng Dongguan, 2009), 29-31.

28 The composer is Xie Chengqiang, a popular composer who has written a number of household pop songs in the 1980s and 1990s, for example, *Xintianyou* and *A True Story*.

All wore make-up and costumes with Chinese elements. The lead singer, Li Zhongqiu, held a *sanxian* with a belt fastening the instrument to his waist so that he could move around onstage and do simple dance movements. Other performers also held Chinese instruments like *sanxian*, *yueqin*, and drums. *Yueqin* and drums were rarely seen in traditional *muyuge* singing. Most of the instruments on stage, including *sanxian*, were used as props, i.e. indispensable signifiers of a traditional Chinese identity. The symbol of tradition was also indicated in the age range of the performers, from primary school students, young women, to the inheritor who was born in the 1940s. The performance signified the transmission and the promising prospect of traditions.

In terms of singing forms, *Evening singing* employed solo, antiphonal, and chorus styles. In contrast to traditional *muyuge* performances, where one single performer sings from the beginning to the end, a chorus is very common in new performances. A chorus plays a significant role in promoting a life-affirming atmosphere and a sense of solidarity. To avoid criticism about distorting the 'authentic' *muyuge* performance, the official adopted a relatively slippery definition to describe the new performances, like '*large-scale muyuge singing and dancing performance*' and *muyuge comedy* when *muyuge* was employed in a comedic drama.[29] The term 'large-scale' indicates the town government's political and economic ambition towards local cultural heritage.

Dongkeng Cantonese, standard Cantonese, and Putonghua

Muyuge is sung in Cantonese, not only in standard Cantonese, but also in Cantonese with accents. Cantonese, widely spoken in the southern provinces of Guangdong and Guangxi, is one of the major dialects of the Chinese language.[30] However, pronunciations and inflections of Cantonese vary in different regions. The accent spoken in Guangzhou is considered as the more

29 *Suno769*, 'Selected Works of Muyuge [木鱼歌作品选辑]', 2014, accessed October 3, 2016, http://dk.suno769.com/subject/myg/zpxj/.

30 Whether Cantonese is a language or a dialect is highly debated. The definition also varies in different cultural contexts. In Hong Kong, Cantonese is juxtaposed with English and Putonghua as an official language. In mainland China, Cantonese is one of the major dialects used in parts of the provinces of Guangdong and Guangxi. Different ways of classifications bear strong political indications. The Cantonese concerned here are within the context of mainland China. For the convenience of discussion, I follow the linguistic system adopted by most mainland linguists, and consider Cantonese as a dialect of the Chinese language. For more details of Chinese dialects, see You Rujie, *Dialectology of the Han Language* (2nd *Edition*) and Maria Kurpaska, *Chinese Language(s)*.

prestigious and standard Cantonese.[31] In the late 1990s, central government began to intensify the promotion of Putonghua (i.e. modern standard Mandarin). Schools are obliged to teach in Putonghua. Cantonese TV stations and radios have been decreasing. Cantonese is in rapid and severe decline. An increasing number of children born in Guangdong Province do not speak and even refuse to learn Cantonese.[32] For many people, even native Cantonese speakers, Cantonese has become a symbol of backwardness and lack of education.[33] Protecting Cantonese is a politically sensitive issue as it is considered a challenge to the dominance of Putonghua and the ideological identity.[34] Given the imminence of standard Cantonese, Cantonese spoken with accents is even beyond public attention and disappearing at a much faster rate. Compared to standard Cantonese, Cantonese with accents is doubly marginalised.

The official reconstruction of *muyuge* carefully avoids political conflicts between Cantonese and Putonghua. Local ICH officials have actively merged local accents of *muyuge* into the dominant narrative of Putonghua. In gala shows and TV documentaries, hosts and narrators always speak in perfect standard Putonghua, regardless of whether the audience are Cantonese speakers or not. In this way, *muyuge* with an 'exotic' local accent is introduced as an 'other' culture, which estranges the local audience from the practice of their own culture. In many *muyuge* performances, Cantonese with local pronunciations is mixed with standard Cantonese and Putonghua. Still, in a few new pieces, e.g. '*Evening Singing*', a Dongkeng Cantonese accent runs throughout the performance. But local accents are not intended to

31 Cantonese, according to You Rujie, has four different sub-dialects among which the Yuehai group (粤海片) has the largest population. It can be difficult for Cantonese speakers from different sub-dialect groups to understand each other. The Yuehai Cantonese speakers mostly come from the Pearl River Delta region, Hong Kong and Macau. The Cantonese spoken among the Yuehai group share similar vocabulary and grammar, but accents vary in different regions in terms of inflections and pronunciations, which is why I use the term 'accents'.

32 Another reason for the prevalence of Putonghua is that a tremendous number of people from other provinces have flooded into Guangdong since economic reform in the late 1970s. However, this reason alone is far from sufficient in explaining the rapid decline of Cantonese since the late 1990s. I agree with Chen Shaojie (2015) that the language policy from central government plays a pivotal role in degrading the position of Cantonese.

33 Shaojie Chen, 'Language Policy and Identity, the Cantonese Culture and Cantonese People since the Economic Reform [语文政策与身份认同：改革开放以来的广府文化与广府人]', *Thinking Hong Kong* (March 2015), accessed October 3, 2016, http://www.thinkinghk.org/archive.

34 In 2010, a member of the Chinese People's Political Consultative Conference (CPPCC) Guangzhou Committee proposed changing the main Cantonese TV channel from Cantonese to Putonghua in order to enhance the 'soft environment' for the 2010 Asian Games. The proposal triggered widespread controversy and a mass protest in Guangzhou.

challenge the dominant Party-state narrative. Instead, it indicated strong solidarity by singing uplifting songs about the dominant ideological values. However, transgressions are inevitable. *Muyuge* is included in the syllabus at a primary and a secondary school in Dongkeng. *Muyuge* class should be the only place in the school where teachers and students speak in Cantonese. Over 40 per cent of the students did not understand Cantonese,[35] but they have to learn local Cantonese in order to complete the course. It is still too early to see how *muyuge* as a primary and secondary school curriculum affects students' perception of Cantonese. But it is a kind of unintended reverse of the mandatory use of Putonghua. Not only did the course teach in local Cantonese, but also require non-Cantonese-speaking students to learn Cantonese. The complicated relations between the Dongkeng Cantonese, standard Cantonese, and Putonghua is also observed in the *muyuge* comedy, *Three radishes in one pit* (三个萝卜一个坑; abbr., *Three radishes*).[36]

Three radishes won the 2010 Star Award, the highest official recognition for folk arts in China.[37] To be precise, it was not a *muyuge* performance, but a theatrical performance featuring several pieces of *muyuge* singing. The Maishen Festival forms the backdrop for a story in which three women compete for a position at a local restaurant to be the official inheritor of a soup made from *yincai* and beef shank, a local dish using the ICH item, *yincai*, a kind of dried radish.

There are two versions of *Three radishes*. In both versions, the professional director and performers, none of whom are native Cantonese speakers, played leading roles in producing the performance. Most of the singing and dialogue were in Putonghua interwoven with pieces of *muyuge* singing. *Muyuge* inheritor Li Zhongqiu wrote the *muyuge* lines and taught performers to sing. The first version was performed at the Maishen Festival's opening ceremony. Except for the Putonghua dialogue, performers sang *muyuge* parts in Dongkeng Cantonese. Li Zhongqiu reminisced that when actors sang the lines in Dongkeng Cantonese, 'the whole square came to life'.[38] Later,

35 Peiyi Huang, personal interview, April 29, 2016.

36 National Digital Cultural Network, 'Quyi: Three Radishes in One Pit [曲艺-三个萝 卜一个坑]', 2013, accessed October 3, 2016, http://www.ndcnc.gov.cn/qunwen/qunxing/201303/t20130312_592746.htm.

37 The Star Award (群星奖) was set up by the Ministry of Culture in 1991, generating over 300 awards every two or three years, covering a wide range of folk cultural practices, from visual arts to performing arts. The award emphasises the participation of non-professional artists and enthusiasts, but it is common that professionals play an essential role during the production and performance process.

38 Zhongqiu Li, personal interview, July 14, 2015.

the town government attempted to push *muyuge* for higher recognition. An official from the Cultural Centre of Dongguan took charge of the new production and worked with another performance group. In the second version, the new production team changed the *muyuge* parts from Dongkeng Cantonese into standard Cantonese.

At the beginning of the performance, four performers arrive on stage to introduce *muyuge*. A woman performer speaks in Cantonese, '*muyuge* sounds amazing, but non-local people might not understand it.' The man performer answers in Cantonese, 'Take your time to guess' (慢慢估). Another woman performer complains, 'What? We need to guess what a performance means?' The third woman performer echoes, 'There are much more non-locals than locals in Dongguan. Such an amazing folk art should also be known to us, the non-locals'. Thus, all performers agree to sing *muyuge* in standard Cantonese and tell stories in Putonghua.

To some extent, the opening dialogue explains the composition of Dongguan's population. According to a statistic in 2013, the total population of Dongguan was 8.31 million with only 1.88 million registered residents, which means nearly 80 per cent of Dongguan residents are fluid population.[39] Many of them are migrant workers from other parts of the country. The characters in the performance consider Cantonese as an oppressive language for the majority of the non-Cantonese residents. This sounds reasonable if isolated from the fact that Cantonese is in rapid decline due to the prevalence of Putonghua. It should also be noted that the revised version was not truly produced for non-Cantonese speakers in Dongguan as the actors claim. In contrast to the popularity of the first version, the second version has rarely been performed since winning the Star Award. In other words, the language of the play was basically revised to please the Putonghua-speaking judges. The alteration from Dongkeng Cantonese to standard Cantonese is another interesting detail. On the one hand, compared with Putonghua, standard Cantonese is considered indigenous enough to expresses a sense of the exotic. On the other hand, standard Cantonese sounds more formal and better educated than local Cantonese. The use of standard Cantonese serves as a compromising solution to demonstrate the sense of indigenousness in a comparatively less 'vulgar' way.

However, the dominance of Putonghua continued throughout the performance. Even some Cantonese-written lines are in a Putonghua style.

39 *Southern Daily*, 'The Top Three in Fluid Population: Dongguan, Shenzhen and Zhongshan [外来人 口比例前三名: 东莞, 深圳, 中山]', *Southern Daily*, December 30, 2014, accessed October 3, 2016, http://gz.southcn.com/g/2014-12/30/content_115335588.htm.

1.2.1 Aunt Zhen (left) and Aunt Chan (right) were singing muyuge in a park in Dongguan, Guangdong Province on 19th October, 2014

Photo by Wen Cuiyan

The title *Three radishes in one pit* is borrowed from a Putonghua idiom 'one radish in one pit', which means one position for one person. The language revision implies that local people might have had limited participation in the production process. In other words, professional actors and artists who can speak standard Putonghua and standard Cantonese were privileged in the official production. The use of Putonghua and standard Cantonese acts as a threshold for local practitioners, especially those of older ages.

In private, cultural officials from Dongkeng stated that they were reluctant to call the second version a *muyuge* performance. Without Dongkeng Cantonese, they stated, the singing lost its authenticity.[40] However, they still gave tacit consent to the revision. Most of the media coverage, therefore, highlighted the performance as *muyuge*. Moreover, the award and related news reports, raised the visibility of *muyuge* together with the Maishen Festival and the aforementioned local soup, which constantly appeared in the performance, thus successfully establishing them as local attractions. It can be argued, therefore,

40 Zhongqiu Li, personal interview, July 14, 2015.

that the tendency to abandon Dongkeng Cantonese has, to a degree, elevated the cultural image of *muyuge* and established it in the dominant discourse.

Conclusion

In this chapter, I have tried to interpret the popular accusation of boredom with regard to traditional performing arts through the close examination of the ICH reconstruction of *muyuge*. In particular, the feeling of boredom indicates a kind of estrangement and exclusion from the traditional cultural practice. Undeniably, the fast-paced modern lifestyle is a key cause of this sense of boredom. What I want to point out is the lurking discourse, which is still shaping people's perception and experience of traditional cultural practices. One of the original intensions of ICH was to challenge the authorised heritage discourse; however, the management and operation of ICH only strengthens this discourse. In China, the operation of ICH is built on governmental management motivated by nationalistic narratives and economic demand. It hinders the ICH campaign from reflecting on the historical causes of the decline of traditional practices and the call for revitalising such practices in recent decades. In the case of *muyuge*, instead of rediscovering its everyday values, the authorities have directly acknowledged the accusation of 'boringness' and made efforts to 'elevate' the genre from ordinariness and 'vulgarity'. Festivalisation, i.e. transferring it to a professional stage performance and reducing the use of local Cantonese are tactics based on the AHD. These tactics aim to extract *muyuge* from the everyday context and turn it into something spectacular.

The story at the beginning of this chapter, however, is not the full story. Although Huang Peiyi did not like the song written by her teacher, she transcribed the lyrics into her notebook (unfortunately, she lost the notebook years later). Her teacher Li Zhongqiu's passion for *muyuge* was recognised by the students. Li later became the ICH inheritor of *muyuge*, but most of his new songs sing in praise of Dongkeng attractions and publicise state policies. The personal connection in heritage transmission became weak. Smith writes, 'the real sense of heritage, the real moment of heritage when our emotions and sense of self are truly engaged, is [...] in the act of passing on and receiving memories and knowledge'.[41] There is much to be done, therefore, between the teacher's enthusiastic teaching and the bored feelings of the students, but efforts should go beyond the limits of the AHD and return to the everyday that people actually live in.

41 Smith, *Uses of Heritage*, 2.

1.2.2 Shi Luguang performed muyuge at home in Dongguan, Guangdong Province in on 14th July 2016

Photo by Wen Cuiyan

1.2.3 Students performing muyuge in an intangible cultural heritage exhibition in Dongguan in July 2015

Photo by Wen Cuiyan

1.2.4 Fang Runzhen (born in 1936) singing muyuge 'Chasing Pheonix in a Buddhist Temple'('禅院追鸾')

http://jeroendekloet.nl/wp-content/uploads/2019/08/1.2.4.-Wen-Cuiyan-Fang-Runzhen-20150821.mp3

Recorded by Wen Cuiyan in Dongguan, Guangdong Province on 21st August,2015.

1.2.5 Blind singer (name unknown) singing muyuge 'Ling Zhaode Selling Water' ('林赵德卖水')

http://jeroendekloet.nl/wp-content/uploads/2019/08/1.2.5.-Wen-Cuiyan-Blindsinger.mp3

Audio bought from a hawker in a grocery market in Dongguan, Guangdong Province

Bibliography

Akagawa, Natsuko. 'Intangible Heritage and Embodiment: Japan's Influence on Global Heritage Discourse'. In *A Companion to Heritage Studies,* eds. William Logan, Mairead Nic Craith, and Ullrich Kockel (Chichester: Wiley Blackwell, 2015), 69-86.

Chen, Shaojie. 'Language Policy and Identity, the Cantonese Culture and Cantonese People since the Economic Reform [语文政策与身份认同：改革开放以来的广府文化与广府人]', *Thinking Hong Kong* (March 2015). Accessed October 3, 2016. http://www.thinkinghk.org/archive.

Dai, Dunfeng. 'China Faces the First Factory Worker Crisis during the Past 20 Years [中国遭遇20年来首次民工荒]', *Southern Weekly*, July 17, 2004. Accessed October 3, 2016. http://www.aisixiang.com/data/3600.html.

Dicks, Bella. *Culture on Display, the Production of Contemporary Visitability* (Maidenhead: Open University Press, 2004).

Dongkeng Annals (Guangzhou: Lingnan Fine Arts Publishing House, 2008).

Harrison, Rodney. *Heritage: Critical Approaches* (London; New York: Routledge, 2013).

He, Duzhong. 'Transmitting *Muyuge*, Five Easy Points, Five Difficult Points and Two Suggestions [传承木鱼歌的五易五难和两点建议]'. In *Selected Works of the Transmission and Preservation of Muyuge* (Dongkeng, China: The Muyuge Transmission Base of Dongkeng Dongguan, 2009), 29-31.

Huang, Wanquan. 'Small Town Dongkeng Hopefully to Become a "Ten-Billion" Town [小镇东坑有望迈进"百亿元"镇街]', *Southern Daily*, January 13, 2016. Accessed October 3, 2016. http://epaper.southcn.com/nfdaily/html/2016-01/13/content_7507974.htm.

Huang, Peiyi. Personal interview. April 29, 2016.

Kurpaska, Maria. *Chinese Language(s), a Look through the Prism of the Great Dictionary of Modern Chinese Dialects* (Berlin; New York: De Gruyter Mouton, 2010).

Li, Zhongqiu. Personal interview. July 14, 2015.

—. 'How to Transmit *Muyuge* [如何做好木鱼歌的传承]'. In *Selected Works of the Transmission and Preservation of Muyuge* (Dongkeng, China: the Muyuge Transmission Base of Dongkeng Dongguan, 2009), 34-5.

Leung, Pui-chee. *Wooden-Fish Books: Critical Essays and an Annotated Catalogue Based on the Collections in The University of Hong Kong* [香港大学所藏木鱼书许录研究] (Hong Kong: Centre of Asian Studies, The University of Hong Kong, 1978).

National Digital Cultural Network. 'Quyi: Three Radishes in One Pit [曲艺-三个萝卜一个坑]', 2013. Accessed October 3, 2016. http://www.ndcnc.gov.cn/qunwen/qunxing/201303/t20130312_592746.htm.

Ng, Shui-hing. 'A Study of the Text of Cantonese Narrative Singing *Muk-jyu-syu*' ('广府话说唱本木鱼书的研究'). PhD dissertation (Hong Kong: The Chinese University of Hong Kong, 1989).

Pinrui Culture. 'Exploration and Transmission: Dongguan *Muyuge* [探寻传承之 东莞木鱼歌]', 2014. Accessed October 3, 2016. http://www.dgswhg.com/fy/sub/?id=135.

Rujie, You. Dialectology of the Han Language (2nd Edition) [汉语言方言学(第二版)] (Shanghai: Shanghai Educational Publishing House, 2016).

Smith, Laurajane. 'Intangible Heritage: A Challenge to the Authorised Heritage Discourse?', *Revista d'etnologia de Catalunya* (2015), 133-42.

—. *Uses of Heritage* (New York: Routledge, 2006).

Southern Daily, 'The Top Three in Fluid Population: Dongguan, Shenzhen and Zhongshan [外来人 口比例前三名: 东莞, 深圳, 中山]'. December 30, 2014. Accessed October 3, 2016. http://gz.southcn.com/g/2014-12/30/content_115335588.htm.

State Council of the People's Republic of China. 'The State Council's Notice about the Fourth Group of National Intangible Cultural Heritage Inventory [国务院关于公布第四批国家级非 物质文化遗产代表性项目名录的通知]', November 11, 2014. Accessed January 23, 2017. http://www.gov.cn/zhengce/content/2014-12/03/content_9286.htm.

—. 'The State Council's Notice about the Third Group of National Intangible Cultural Heritage Inventory [国务院关于公布第三批国家级非物质文化遗产代表性项目名录的通知]', May 23, 2011. Accessed January 23, 2017. http://www.gov.cn/zwgk/2011-06/09/content_1880635.htm.

—. 'The State Council's Notice about the Second Group of National Intangible Cultural Heritage Inventory and the Expanded Items of the First Group of National Intangible Cultural Heritage Inventory [国务院关于公布第二批国家级非物质 文化遗产名录和第一批国家级非物质文化遗产扩展项目名录的通知]', June 7, 2008. Accessed January 23, 2017. http://www.gov.cn/zwgk/2008-06/14/content_1016331.htm.

—. 'The State Council's Notice about the First Group of National Intangible Cultural Heritage Inventory [国务院关于公布第一批国家级非物质文化遗产名录的通知]', May 20, 2006. Accessed January 23, 2017. http://www.gov.cn/zhengce/content/2008-03/28/content_5917.htm.

—. 'The Office of the State Council's Opinions about Strengthening the Safeguarding of China's Intangible Cultural Heritage [国务院办公厅关于加强我国非物 质文化遗产保护工作的意见]', March 26, 2005. Accessed January 23, 2017. http://www.gov.cn/gongbao/content/2005/content_63227.htm.

Sun0769, 'Three Ancient Cultures Boost Agricultural Park and Create New "Selling Points" of Leisure Tourism ["三古"文化助推农业园打造休闲旅游新"卖点"]'. December 17, 2015. Accessed October 3, 2016. http://dk.sun0769.com/jinrikandian/jinriyaowen/59988.shtml?1450342345.

—. 'Selected Works of *Muyuge* [木鱼歌作品选辑]'. 2014. Accessed October 3, 2016. http://dk.sun0769.com/subject/myg/zpxj/.

—. 'The Representative Inheritor of *Muyuge* Li Zhongqiu [木鱼歌代表性传承人 李仲球]'. May 27, 2014. Accessed October 3, 2016. http://dk.sun0769.com/subject/myg/fydbxccr/49233.shtml.

UNESCO. 'Text of the Convention for the Safeguarding of the Intangible Cultural Heritage', 2003. Accessed October 3, 2016. http://www.unesco.org/culture/ich/en/convention.

Wang, Jing. 'Culture as Leisure and Culture as Capital', *Positions: East Asia Cultures Critique* 9.1 (2001): 69-104.

Xiang, Kailai. 'From the Wave of Migrant Workers to the Crisis of Migrant Workers, What Does it Indicate? ["民工潮"到"民工荒", 说明了什么?]', *Xinhua News*, September 6, 2004. Accessed October 3, 2016. http://news.xinhuanet.com/focus/2004-09/06/content_1938673.htm.

Yang, Baolin. Interview by letter. July 20, 2015.

—. 'The Study of Dongguan *Muyuge* [东莞木鱼歌研究]', *Journal of Dongguan University of Technology* (*Social Science Edition*) 12.2 (2005).

Yung, Bell. *Cantonese Opera Performance as Creative Process* (Cambridge: Cambridge University Press, 1989).

Zhang, Tiewen. 'Exploration and Transmission: Dongguan *Muyuge* [探寻传承之东莞木鱼歌]'. Personal interview by Pinrui Culture. 2014. Accessed October 3, 2016. http://www.dgswhg.com/fy/sub/?id=135.

About the Author

Wen Cuiyan is a lecturer at the Department of Literature and Media in Nanfang College of Sun Yat-Sen University. She holds her PhD in Cultural Studies from Lingnan University, Hong Kong. Her research interests include everyday life, critical heritage studies, community arts, and traditional performing arts.

1.3 You Must (Not) Be Bored!

Boredom and Creativity in Global Capitalism

Esther Peeren

Abstract
Boredom and creativity are generally considered mutually exclusive; yet, it is possible to imagine forms of creative boredom or boring creativity. Whether the two notions are compatible or clash depends on their historically and culturally specific meanings. This chapter begins by tracing how Michel de Certeau, Luce Giard, and Jonathan Crary position boredom and creativity in the context of capitalism's increasingly firm imperative to always be doing something. Next, Peeren zooms in on the claims to a boredom-busting creativity made by two popular cultural objects with global reach: the Bored Panda website and adult colouring books. She argues that, besides reinforcing the directive to keep busy, these claims also propose alternative conceptions of creativity and boredom with critical potential.

Keywords: boredom, creativity, capitalism, unnameable, Bored Panda, adult colouring books

Boredom and creativity are generally considered mutually exclusive, yet it is not impossible to imagine forms of creative boredom or of bored or boring creativity. Whether the two notions are compatible or clash is dependent on their historically and culturally specific meanings. This short intervention begins by tracing how Michel de Certeau, Luce Giard, and Jonathan Crary position boredom and creativity in the context of capitalism's increasingly firm imperative to always be doing something. Next, I zoom in on the claims

Kloet, Jeroen de, Chow Yiu Fai, and Lena Scheen (eds), *Boredom,* Shanzhai, *and Digitisation in the Time of Creative China*. Taylor & Francis Group, 2019
DOI: 10.5117/9789462984745_CH1.3

to a boredom-busting form of creativity made by two popular contemporary cultural objects with global reach: the Bored Panda website and adult colouring books. I argue that such claims not only reinforce the directive to keep busy but can also be seen to propose alternative conceptions of creativity and boredom with critical potential.

In *The Practice of Everyday Life*, first published in 1984, Michel de Certeau writes:

> In our society, the absence of work is non-sense; it is necessary to eliminate it in order for the discourse that tirelessly articulates tasks and constructs the Occidental story of 'There's always something to do' to continue.[1]

Not working is senseless within a capitalist system characterised by a relentless push to act in an economically productive manner. Consequently, the dying – who can no longer work – and the lazy – who refuse to work at their full capacity or at all – are rendered unthinkable and 'unnameable'.[2] To these two groups may be added the bored, who, by claiming that there is nothing to do or at least nothing of interest to them, also fall outside the prevailing discourse. However, the bored challenge this discourse in a way the dying and the lazy cannot. Whereas the dying and the lazy do not participate because they are, respectively, unable or unwilling to exert themselves, the bored, in their 'refusal of engagement',[3] deny the worth of what there is to do, proclaiming it inadequate and unfulfilling. To find something boring is always to devalue and critique it.

But boredom is never only a critical judgement on something or someone else. Those who feel bored are also negatively affected themselves, as boredom is generally experienced as an undesirable, frustrating state, especially when the (lack of) impulse that produces it cannot easily be escaped. This burden of boredom is carried not only by those who, by economic necessity, engage in the many forms of repetitive, routinised labour on which global capitalism continues to rely, but also by those whose work has been lightened by the gains of labour movements and technological advances. As Luce Giard writes in the 1998 edition of *The Practice of Everyday Life, Volume 2: Living and Cooking*, the mechanisation of domestic kitchens

1 Michel de Certeau, *The Practice of Everyday Life* (Berkeley, CA: University of California Press, 1988), 191.

2 Ibid.

3 Patricia Meyer Spacks, *Boredom: The Literary History of a State of Mind* (Chicago, IL; London: University of Chicago Press, 1996), 2.

over the course of the twentieth century 'allowed people to save time and decrease one type of fatigue, but this was done in order to give rise to gray, homogeneous, empty time, the time of effortless and joyless boredom'.[4] Here, boredom results from manual labour being replaced by machines only requiring minimal input from the cook, who is mostly reduced to waiting. In contrast to the active, critical boredom that identifies capitalism's story of 'there's always something to do' as insufficient, this is an 'effortless and joyless' boredom unassuaged by the distraction of physical exhaustion. It is a passive boredom without content, born of a feeling of superfluity, of finding oneself expendable.

The 'empty time' of this joyless boredom is hostile to creativity in the everyday sense of the 'modest inventiveness' that, through 'ephemeral results' such as a hand-cooked meal, allows the definition of a distinctive '*lifestyle*'.[5]. This creativity cannot be recovered by replicating cooking practices of the past through an 'archaistic nostalgia', while 'frenetic overmodernisation' would further erode it.[6] Where Giard sees it persisting is in certain localised 'microexperiences' that 'try modestly to invent other behaviours' – behaviours that, in combining the old (handwork) with the new (machine work), 'define a lifestyle straddling the two cultures and their two temporalities', countering the dominant tendency towards standardisation.[7]

In the 21st century, the capitalist discourse of 'there's always something to do' has accelerated and spread. No longer a specifically Occidental story, it has found global dominance. Moreover, the growing emphasis on consumption and commodification has added the imperative of 'there's always something to buy'. Jonathan Crary, in *24/7: Late Capitalism and the Ends of Sleep*, paints a nightmarish picture of a global 'non-stop life-world' that 'has the semblance of a social world, but [...] is actually a non-social model of machinic performance and a suspension of living that does not disclose the human cost required to sustain its effectiveness'.[8] The post-historical temporality of this world resembles Giard's empty time: Crary describes it as a machinic time hostile to 'any long-term undertakings' or 'fantasies of "progress" or development', not just at the individual level but also at the

4 Luce Giard, 'Part II: Doing Cooking', in *The Practice of Everyday Life, Volume 2: Living and Cooking*, eds. Michel de Certeau, Luce Giard and Pierre Mayol (Minneapolis, MN; London: University of Minnesota Press, 1998), 212.

5 Ibid., 213. Emphasis in original.

6 Ibid.

7 Ibid.

8 Jonathan Crary, *24/7: Late Capitalism and the Ends of Sleep* (London; New York: Verso, 2014), 8-9.

collective one.[9] Its space is fully illuminated, dispersing the 'shadows', 'obscurity', and 'alternate temporalities' in which Giard's microexperiences could still take shape.[10] Arguing that 'the highest premium is placed on activity for its own sake', Crary quotes Luc Boltanski and Eve Chiapello's updated, neo-liberal version of de Certeau's story of capitalism:

> To always be doing something, to move, to change – this is what enjoys prestige, as against stability, which is often synonymous with inaction.[11]

Reconfigured as inertia, stability is the new unthinkable and unnameable. This represents an amplification of the story presented by de Certeau because it renders non-sensical not just the absence of work, but also its stable presence, excluding, in addition to the dying, the lazy, and the bored, the satisfied – those contents to stay in place.

Crary locates a possible disruption of the 24/7 global economy in the persistence of a particular form of inaction, namely the 'scandal of sleep'.[12] He sees sleep as offending the command 'to always be doing something' since it 'poses the idea of a human need and interval of time that cannot be colonised and harnessed to a massive engine of profitability'.[13] This ignores that sleep is both intensely commodified – the global sleep aids market is estimated to be worth $80,814.7 million by 2020[14] – and increasingly considered essential to enhancing productivity.[15] Redefined as an action that underpins the non-stop economy, sleeping has become part of the 'there's always something to do'.

Similarly, in a 24/7 world that values movement and change above all else, boredom is no longer a potential challenge to the capitalist system or

9 Ibid., 9.
10 Ibid., 19.
11 Ibid., 15.
12 Ibid., 11.
13 Ibid., 10-11.
14 This is a projected valuation according to a 2015 study by Persistence Market Research. See: http://www.persistencemarketresearch.com/market-research/sleep-aids-market.asp.
15 Gregory Belenky, Nancy J. Wesensten, David R. Thorne, Maria L. Thomas, Helen C. Sing, Daniel P. Redmond, Michael B. Russo, and Thomas J. Balkin, 'Patterns of Performance Degradation and Restoration during Sleep Restriction and Subsequent Recovery: A Sleep Dose-response Study', *Journal of sleep research* 12.1 (2003): 1-12; Meagan Daley, Charles M. Morin, Mélanie LeBlanc, Jean-Pierre Grégoire, Josée Savard, and Lucie Baillargeon. 'Insomnia and Its Relationship to Health-care Utilization, Work Absenteeism, Productivity and Accidents', *Sleep medicine* 10.4 (2009): 427-438; Mark R. Rosekind, Kevin B. Gregory, Melissa M. Mallis, Summer L. Brandt, Brian Seal, and Debra Lerner. 'The Cost of Poor Sleep: Workplace Productivity Loss and Associated Costs', *Journal of Occupational and Environmental Medicine* 52.1 (2010): 91-98.

a pernicious side-effect that may be countered through microexperiences, but what drives the system, albeit in a paradoxical way. On the one hand, boredom never has to occur, since there are now so many things to do all the time that it is impossible to claim not to be engaged by anything. On the other hand, boredom has to be incessantly conjured in order to push people into constant action and, above all, consumption. Thus, within the current system of global capitalism, boredom simultaneously makes no sense and total sense. There is no time for it and there is no time when it is not looming. Boredom has become eminently escapable – the empty time produced by kitchen appliances' efficiency is now easily filled by turning to the instant entertainment offered by a multitude of digital devices – while at the same time being necessary to produce non-stop pressure to move on to the next thing. Just as the new adage with regard to sleeping is not, as Crary maintains, 'You shall not sleep!', but rather 'You shall sleep enough to render you optimally productive as a worker and consumer!', the new adage with regard to boredom is not 'You shall not be bored!' but the maddeningly contradictory 'You must (not) be bored!'

How to put such an illogical command into practice is demonstrated by the Bored Panda website, which features a wide array of brief articles, lists, and image galleries designed to dispel boredom, while relying on the constant renewal of boredom to ensure traffic to and within the site.[16] Its perplexing name – is it asserting that pandas are particularly prone to boredom (if so, how was this ascertained?), or is the panda, as the epitome of animal cuteness, supposed to be seen as a guaranteed antidote to boredom? – underlines the lack of a logic for the site's disparate content, other than it being so varied that, as a whole, it could never be deemed boring. Significantly, Bored Panda presents itself as a 'leading art, design and photography community for creative people' and relies on people uploading amusing content of their own making. Here, boredom appears as the 'creative impetus' it was often seen as in the nineteenth century,[17] but in a twisted manner. It is not one's own sense of boredom that prompts a spurt of creativity but that of other people, while the product of creativity is not valued aesthetically but solely in terms of whether it draws and engages the attention of the nominally bored.

In what sense, then, is the site's content creative? The site states that 'anyone can write on Bored Panda', so there is no assessment – other than through the number of times a post is clicked on, commented on,

16 See www.boredpanda.com.

17 Spacks, *Boredom*, 3. See also Lindner in this volume.

or shared on social media – of the degree of creativity entries must have to be able to dispel boredom. Most posts are certainly not creative in the globally dominant sense of the term, heralding from romanticism and associating it with originality, beauty, and artistic genius. It seems as if the aim of dispelling boredom, which, 'unlike its more dignified cousin ennui [...] is often considered a trivial emotion',[18] only requires a mundane form of creativity, especially when the site's commodification of the relation between boredom and creativity relies on users moving quickly from post to post rather than being arrested by a single one. However, while Bored Panda is undoubtedly complicit with global capitalism in urging both its visitors and its creative community 'to always be doing something', the site's light-hearted absurdity also prompt a questioning of this command, not least because it is likely to be used during work-time as a distraction from productivity.[19] In addition, the broad, modest notion of creativity adopted by Bored Panda, besides being a ploy to generate free content, enables those who would in other contexts not be recognized as creative to show off what they have made, to mostly supportive comments. Thus, Bored Panda may facilitate meaningful microexperiences in Giard's sense.

Like Bored Panda, adult colouring books, currently enjoying bestseller status across the world, are marketed as combating boredom and stress through creativity.[20] While the idea of fighting stress, associated with overstimulation, by engaging in an activity that for most adults requires minimal effort and skill makes sense, it seems paradoxical to use this same activity to battle boredom, often considered to result precisely from repetitive, unchallenging tasks. However, it could be argued that what these colouring books challenge is a different type of boredom, namely the erratic boredom generated by the ever-shortening attention spans of those trying to comply with neo-liberalism's demand for continuous change and novelty. Adult colouring books then appear as part of a larger

18 Spacks, *Boredom*, xi.

19 There is now also research suggesting that looking at cute images may actually enhance productivity: Nittono et al. found that 'participants performed tasks requiring focused attention more carefully after viewing cute images'. Still, only part of Bored Panda's content fits the cuteness category.

20 Adrienne Raphel, 'Why Adults Are Buying Coloring Books (for Themselves)', *The New Yorker*, July 12, 2015, accessed November 4, 2016, http://www.newyorker.com/business/currency/why-adults-are-buying-coloring-books-for-themselves; Kate Harrison, 'The Adult Coloring Craze Continues And There Is No End In Sight', *Forbes*, February 2, 2016, accessed November 4, 2016, http://www.forbes.com/sites/kateharrison/2016/02/02/the-adult-coloring-craze-continues-and-there-is-no-end-in-sight/#2519e7ba643b.

movement appealing to slowness – from slow food and slow fashion to slow thinking[21] and slow professors[22] – in the face of the seemingly limitless acceleration of the 24/7 world. Carefully colouring within the lines challenges the fetishisation of newness, of having to always think outside the box, in a reversal of adult colouring books from the 1960s, such as the *Executive Coloring Book*, which, as part of that era's counterculture, took aim at the conformity its form of capitalism enforced by depicting it as a soul-destroying colouring within the lines and utter blandness: 'This is my suit. Colour it gray or I will lose my job'.[23] Crucially, these colouring books tended not to be coloured in, as doing so would mean to conform rather than to critique the system.[24]

It is almost impossible to withstand the urge to dismiss adult colouring books as boring, immature, or infantilising, and devoid of any creative or artistic value: in *The New Yorker*, the craze is seen as part of a 'Peter Pan market',[25] while a blogger refers to 'poorly executed cheater art'.[26] Valid critiques can indeed be made of the hyperbolic claims to colouring's therapeutic and spiritual value, as well as of the genre's cultural appropriation of the Hindu, Buddhist, and Jainist mandala symbol. Yet, it is important to realise that dismissing adult colouring as not 'really' creative because it supposedly does not offer any room for originality reifies a particular, Western notion of creativity and echoes the disparagement of Chinese *shanzhai* or 'fake' art.[27] What is ignored is not only that originality may be located in many aspects besides the delineation of an image (in colour choice and combination, pencil stroke, thickness of colour application, etc.), but also that creativity is not necessarily predicated on originality; skill, acquired through laborious practice and the copying of existing works, may also found and enhance creativity, just like creativity research has shown

21 Daniel Kahneman, *Thinking, Fast and Slow* (New York: Farrar, Strauss and Giroux, 2011).

22 Maggie Berg and Barbara Seeber, *Slow Professor: Challenging the Culture of Speed in the Academy* (Toronto: University of Toronto Press, 2016).

23 Quoted in Laura Marsh, 'The Radical History of 1960s Adult Coloring Books', *New Republic*, December 28, 2015, accessed November 4, 2016, https://newrepublic.com/article/126580/radical-history-1960s-adult-coloring-books.

24 Ibid.

25 Raphel, 'Why Adults Are Buying Coloring Books'.

26 Joni Edelman, '7 Reasons Adult Coloring Books Are Ridiculous', *Ravishly*, October 7, 2015, accessed November 4, 2016, http://www.ravishly.com/2015/10/07/7-reasons-adult-coloring-books-are-ridiculous.

27 Winnie Wong, *Van Gogh on Demand: China and the Readymade* (Chicago, IL: University of Chicago Press, 2014).

boredom can do.[28] Present-day adult colouring books may indeed offer a 'neat package of therapy, escape and nostalgia'[29] that keeps people engaged in consumption and offers them a no-risk, low-effort sense of achievement, but they also encourage the kind of absorbing, durable activity that escapes global capitalism's imperative of 'you must (not) be bored!' In addition, like the Bored Panda website, these colouring books endorse an alternative, more inclusive notion of creativity that allows activities often seen as unproductive, valueless, and boring to make sense and to be endorsed, as when people showcase the results of their colouring on social media.

Bibliography

Belenky, Gregory, Nancy J. Wesensten, David R. Thorne, Maria L. Thomas, Helen C. Sing, Daniel P. Redmond, Michael B. Russo, and Thomas J. Balkin. 'Patterns of Performance Degradation and Restoration during Sleep Restriction and Subsequent Recovery: A Sleep Dose-response Study', *Journal of sleep research* 12.1 (2003): 1-12.

Berg, Maggie, and Barbara Seeber. *Slow Professor: Challenging the Culture of Speed in the Academy* (Toronto: University of Toronto Press, 2016).

Crary, Jonathan. *24/7: Late Capitalism and the Ends of Sleep* (London; New York: Verso, 2014).

Daley, Meagan, Charles M. Morin, Mélanie LeBlanc, Jean-Pierre Grégoire, Josée Savard, and Lucie Baillargeon. 'Insomnia and Its Relationship to Health-care Utilization, Work Absenteeism, Productivity and Accidents', *Sleep medicine* 10.4 (2009): 427-438.

de Certeau, Michel. *The Practice of Everyday Life* (Berkeley, CA: University of California Press, 1988).

Edelman, Joni. '7 Reasons Adult Coloring Books Are Ridiculous', *Ravishly*, October 7, 2015. Accessed November 4, 2016. http://www.ravishly.com/2015/10/07/7-reasons-adult-coloring-books-are-ridiculous.

Giard, Luce. 'Part II: Doing Cooking'. In *The Practice of Everyday Life, Volume 2: Living and Cooking*, eds. Michel de Certeau, Luce Giard and Pierre Mayol (Minneapolis, MN; London: University of Minnesota Press, 1998), 149-248.

28 Sandi Mann and Rebekah Cadman, 'Does Being Bored Make Us More Creative?', *Creativity Research Journal* 26.2 (2014): 165-73; see also Wilf in this volume.
29 Laura Marsh, 'The Radical History of 1960s Adult Coloring Books', *New Republic*, December 28, 2015, accessed November 4, 2016, https://newrepublic.com/article/126580/radical-history-1960s-adult-coloring-books.

Harrison, Kate. 'The Adult Coloring Craze Continues and There Is No End in Sight', *Forbes*, February 2, 2016. Accessed November 4, 2016. http://www.forbes.com/sites/kateharrison/2016/02/02/the-adult-coloring-craze-continues-and-there-is-no-end-in-sight/#2519e7ba643b.

Kahneman, Daniel. *Thinking, Fast and Slow* (New York: Farrar, Strauss and Giroux, 2011).

Mann, Sandi, and Rebekah Cadman. 'Does Being Bored Make Us More Creative?', *Creativity Research Journal* 26.2 (2014): 165-73.

Marsh, Laura. 'The Radical History of 1960s Adult Coloring Books', *New Republic*, December 28, 2015. Accessed November 4, 2016. https://newrepublic.com/article/126580/radical-history-1960s-adult-coloring-books.

Nittono, Hiroshi, Michiko Fukushima, Akihiro Yano, and Hiroki Moriya. 'The Power of Kawaii: Viewing Cute Images Promotes a Careful Behaviour and Narrows Attentional Focus', *PloS one* 7.9 (2012): e46362.

Raphel, Adrienne. 'Why Adults Are Buying Coloring Books (for Themselves)', *The New Yorker*, July 12, 2015. Accessed November 4, 2016. http://www.newyorker.com/business/currency/why-adults-are-buying-coloring-books-for-themselves.

Rosekind, Mark R., Kevin B. Gregory, Melissa M. Mallis, Summer L. Brandt, Brian Seal, and Debra Lerner. 'The Cost of Poor Sleep: Workplace Productivity Loss and Associated Costs', *Journal of Occupational and Environmental Medicine* 52.1 (2010): 91-98.

Spacks, Patricia Meyer. *Boredom: The Literary History of a State of Mind* (Chicago, IL; London: University of Chicago Press, 1996).

Wong, Winnie Won-yin. *Van Gogh on Demand: China and the Readymade* (Chicago, IL: University of Chicago Press, 2014).

About the Author

Esther Peeren is Professor of Cultural Analysis at the University Amsterdam, Director of the Amsterdam School for Cultural Analysis (ASCA) and Principal Investigator of the ERC-funded project Imagining the Rural in a Globalizing World (2018-2023). Recent publications include *The Spectral Metaphor: Living Ghosts and the Agency of Invisibility* (Palgrave, 2014) and the edited volumes *Peripheral Visions in the Globalizing Present* (Brill, 2016, with Hanneke Stuit and Astrid Van Weyenberg) and *Global Cultures of Contestation* (Palgrave, 2018, with Robin Celikates, Jeroen de Kloet and Thomas Poell).

1.4 Boredom and Creativity in the Era of Accelerated Living

Christoph Lindner

Abstract

Rethinking the historical relation between boredom and creativity in the era of accelerated living, this chapters examines the impact of neoliberal globalisation's culture of speed and connectivity on creative practice. It argues that boredom today has generally shifted from being an affected aesthetic pose involving stillness and retreat to become more of an involuntary response to the exhausting hyperactivity and excessive production characterising contemporary life.

Keywords: globalisation, contemporary art, creativity, cities, China

I took the photo below at the Kowloon campus of Hong Kong Baptist University. I was pleasantly surprised to find this small effort at urban wall gardening clinging to the side of a high-rise building. Part guerrilla garden and part art installation, the urban wall garden struck me as a creative intervention in the space and fabric of the built environment. It repurposes plastic waste by using empty bottles as plant pots. It brings a splash of life and colour to a dull, monochromatic space. It communicates care and attention for an overlooked non-place. It uses the aesthetics and materiality of green insurgency to gently push back – at an intimate scale – against the rampant denaturalisation of the city. And unlike some forms of urban art/intervention that rely on tactics of speed, mobility, and transience (such as flash mobs or parkour) the urban wall garden involves emplacement, stillness, duration. But vertical gardening, as it is more widely called, is not a practice unique to Hong Kong. Rather, it is one that circulates transnationally, materialising on urban walls throughout the world, largely thanks to the global ubiquity of plastic bottling (which is a different discussion about

Kloet, Jeroen de, Chow Yiu Fai, and Lena Scheen (eds), *Boredom,* Shanzhai, *and Digitisation in the Time of Creative China.* Taylor & Francis Group, 2019
DOI: 10.5117/9789462984745_CH1.4

1.4.1 An urban wall garden on the side of a high-rise building at Hong Kong Baptist University

waste and petroculture), as well as to the highly developed networks of communication and knowledge sharing that characterise contemporary activism, street art, and urban farming movements.

The image itself was casually snapped with my smart phone in a moment of distraction and automatically uploaded to social media, where it was picked up and globally dispersed through the techno-informational 'space of flows', to borrow Manuel Castells' phrase.[1] I reproduce the image here not because of any potential aesthetic merit, but because both the object itself (the hypermediated digital image) and the slow creative practice it depicts (vertical gardening) bring together my main concerns in this essay. In what follows, I draw on the Romantic view of creativity as epiphany to question the impact of neoliberal globalisation's culture of speed and connectivity on creative practice.

Tranquillity

Western culture has a long tradition of associating creativity with spontaneity, as mythologised in the ancient Greek tale of Archimedes' mathematical bathtub epiphany. It was not until the rise of Romanticism in the late eighteenth century, however, that the insight and innovation associated with the 'Eureka effect' in

1 Manuel Castells, *The Rise of the Network Society* (Oxford: Blackwell, 2000), 409.

science[2] also became tied to artistic production and, more broadly, the field of aesthetics.[3] Central to this development were writers such as the British poets Samuel Taylor Coleridge and William Wordsworth, who carefully crafted a public image of themselves as solitary poet-geniuses prone to sudden and profound creative revelations when confronted by truth or beauty – usually in the form of a natural landscape, a rustic human figure, or an architectural ruin.

In the 1800 preface to *Lyrical Ballads*, the joint collection of poems first published with Coleridge in 1798, Wordsworth famously articulates his theory of creativity. 'Poetry', he writes, 'is the spontaneous overflow of powerful feelings' and 'takes its origin from emotion recollected in tranquillity'.[4] This statement is significant because it understands artistic creation as involving both impulse and planning, both stimulation and boredom (if we think of boredom as the endured passage of time). Central to Wordsworth's theory is that, while art derives from intense emotional experience, the act of aesthetic creation itself necessitates undisturbed time (tranquillity) during which the memory of strong emotions can be carefully conjured up and imaginatively intensified, before being filtered through the artist's aesthetic sensibility and redirected into the material form of the work of art – in this case, poetry.

In effect, Wordsworth formulates a cliché that has dominated popular perceptions of artistic production for many centuries: namely, the image of the artist as a tortured soul toiling in solitude under extreme mental pressure and subject to sudden eruptions of creativity. Whether accurate or not, it is an image that has accompanied the mythos surrounding many iconic writers, visual artists, designers, architects, musicians, and more. One of the problems is that, in order to locate creativity in the singular mind of the artist, the Romantic self-image of creativity constructed by writers like Wordsworth, and amplified by contemporaries such as Blake, Byron, and De Quincey, underplays collaboration, influence, and worldly engagement. In this version, creativity occurs under highly privileged and protected conditions of seclusion – at a remove from society, interaction, exchange. And yet, as Wordsworth's own co-publishing with Coleridge demonstrates, creative work is frequently – even necessarily – collaborative in nature, in the sense that, like language itself, ideas do not materialise out of nowhere but are formed by interaction with and influence from other ideas and, by extension, other people.

2 David N. Perkins, *The Eureka Effect: The Art and Logic of Breakthrough Thinking* (New York: W.W. Norton, 2001).

3 See also Wilf in this volume.

4 William Wordsworth, 'The Preface to Lyrical Ballads'. In *Lyrical Ballads and Other Poems* (Ware: Wordsworth Editions, 2003), 21.

I am reminded here of Roland Barthes' wonderfully provocative claim in his 1967 essay, 'The Death of the Author', that 'the text is a tissue of quotations drawn from the innumerable centres of culture [...] the writer can only imitate a gesture that is always anterior, never original'.[5] The idea that creative expression is always already unoriginal to some degree – that its meaning resides in its relation to the meaning of other texts and is, in this sense, derivative or gestural – can be widely applied beyond literature to other art forms. More to the point, Barthes' theory, partly articulated in the title of his essay, rejects the cult of authorship promoted by writers like Wordsworth in favour of an understanding of creative production as a networked, relational activity.

Clearly, I find Barthes' poststructuralist spin on creative production more convincing than Wordsworth's Romantic meditation on spontaneous overflows. My interest, however, is not in arguing the merits of these positions, nor even in developing some working definition of creativity. Rather, I want to question the widespread popular view – still dominant today – that creativity is constituted by sudden flashes of imaginative insight. Such an understanding of creativity does more than obscure its interactive and collaborative dimensions. It renders the labour of creative work invisible. It privileges the artist as a singularity. It fetishises originality. And it neglects the socio-economic realities, as well as the cultural politics and technological-material conditions, that shape (and are shaped by) creative practice.

Precarity

Fast forward to the 21st century and to the present era of globalisation, accelerated urbanism, transnational mobility, and digital nativism. Yes, the Romantic cliché of the lone creative genius endures. In the Chinese context, for example, it can be seen in the international superstardom of the artist Ai Weiwei. But if we look beyond the elite exceptions of today's global art world, the broader picture that emerges is very different. For one thing, creativity has long ceased to belong to the domain of aesthetics. Following the industrial turn and the rise of modern capitalism, the professionalisation of the artist in the nineteenth century (including the commodification of art itself) paved the way for what the scholar-consultant Richard Florida

5 Roland Barthes, 'The Death of the Author'. In *Image-Music-Text*, ed. and trans. Stephen Heath (New York: Hill and Wang, 1977), 146.

has described as the 'rise of the creative class'[6] in the late-twentieth and early-twenty-first centuries.

Although heavily critiqued for oversimplifying the complexities of urban living and for over-privileging economic development as a marker of cultural vibrancy, Florida's theory that we are living in the 'creative age' has gained considerable traction in urban and cultural policy worldwide and has been actively mobilised by countless municipalities seeking post-industrial revitalisation. Such efforts to reinvent declining cities as creative cities frequently fail or end up exacerbating gentrification, inequality, and segregation, as Florida himself has been forced to concede in later work.[7] Partly tapping into the zeitgeist of hipsters, laptops, and cafés, and partly contributing to that zeitgeist by promoting professionalised forms of creativity as instruments of urban prosperity, Florida's vision of urban renewal promotes an understanding of creativity as a quotidian, workplace activity taking place in the wider context of a global market economy and within the loose clustering of output-oriented, hype-driven, tech-centred professional fields we call the 'global creative industries'.[8]

One of the dark sides of the creative turn in urban policy is the mobilisation of the artist as a gentrifier. This typically involves attracting creative professionals to 'ailing' areas of a city so that their presence and activity can create a positive cultural-economic vibe that attracts rapid development. In this scenario, artists are lured by the affordability of space only to become displaced, along with other low-income residents, once the neighbourhood 'renews' and property prices increase.[9] Key to this phenomenon is the widespread precarity of creative work, which not only makes artists and other creative professionals susceptible to policy-led neoliberal renewal in the first place, but also ensures that the cycle continues.

As these dynamics suggest, the Romantic ideal of aesthetic innovation, and in particular Wordsworth's call for tranquillity and retreat, is increasingly incompatible with the everyday realities shaping contemporary life. As Jonathan Crary argues, for example, late capitalism has given rise to an exhausting war on rest, in which the forms of disengagement needed

6 Richard Florida, *The Rise of the Creative Class* (New York: Basic Books, 2002).

7 Florida, *The New Urban Crisis: How Our Cities Are Increasing Inequality, Deepening Segregation, and Failing the Middle Class – and What We Can Do About It* (New York: Basic Books, 2017).

8 Terry Flew, *Global Creative Industries* (Cambridge: Polity, 2013).

9 Loretta Lees, Hyun Bang Shin, and Ernesto Lopez-Morales, *Planetary Gentrification* (Cambridge: Polity, 2016); Sharon Zukin, *Naked City: The Death and Life of Authentic Urban Places* (New York: Oxford University Press, 2011).

to experience tranquillity are under relentless attack.[10] 'The expanding, non-stop life-world of 21st-century capitalism', he observes, is marked by a constant struggle between 'shifting configurations of sleep and waking, illumination and darkness, justice and terror', in which human life is inscribed 'into duration without breaks, defined by a principle of continuous functioning'.[11]

Many mourn the loss of tranquillity brought about by neoliberal globalisation's 'culture of speed',[12] as well as the loss of the privileged social and material conditions that enable such states of repose. One reaction against accelerated living can be found in the slow living movements that have proliferated worldwide, beginning with Slow Food in Italy in the 1980s and now encompassing a broad spectrum of slow movements extending from art and design to science and finance.[13] What unites the global slow living trend is the view that slowness – as both an embodied practice and an affective condition – can be used as 'a strategy for confronting globalisation, neoliberal, and the associated accelerations of everyday life, transport, communication, and economic exchange'.[14] I do not wish to discount the real and potential benefits of slow movements to individuals or communities. Yet the critique can be made that slowness is frequently reactive and exclusive – too often an expensive lifestyle choice that temporarily mitigates the effects of accelerated living rather than delivering a genuine, long-term, systematic alternative.

This tension is perfectly captured in an advertising poster I once encountered inside a trendy organic 'farm-to-table' burger restaurant in Berkeley, California. Overlaying images of a chef cooking, the poster promotes the restaurant as '(slow) fast food', effectively acknowledging the vicarious form that the slowness of organic burgers assumes in this rapid-dining context. From the local sourcing of food to its preparation and presentation, the restaurant engages in slowness (or at least the performance of it) on behalf of rushing customers who do not have the time or desire to slow down themselves. In scenarios such as these, as I had to admit to myself while consuming my slow-fast meal, slowness is the lie we tell ourselves so that we can continue to indulge the neoliberal dream of accelerated

10 Jonathan Crary, *24/7: Late Capitalism and the Ends of Sleep* (New York: Verso, 2013). See also Peeren in this volume.

11 Crary, *24/7*, 8.

12 John Tomlinson, *The Culture of Speed: The Coming of Immediacy* (London: Sage, 2007).

13 Carl Honoré, *In Praise of Slow: Challenging the Cult of Speed* (New York: Harper Collins, 2004).

14 Christoph Lindner and Miriam Meissner, 'Slow Art in the Creative City: Amsterdam, Street Photography, and Urban Renewal', *Space and Culture* 18.1 (2015): 4.

1.4.2 A poster at an organic 'farm-to-table' burger restaurant in Berkeley, California

living, complete with all its inequalities, excesses, and contradictions. Another way of thinking about such contemporary contradictions is to say that slow is another dimension of fast – that slowness often depends on hidden forms of speed and ultimately helps us to endure the growing intensity of 24/7 living, making the very condition of acceleration all the more viable.

It is against this backdrop of neoliberal globalisation and accelerated living that I wish to return to the topic of boredom and creativity. The reason is that boredom today, especially in the context of creative practice, has generally shifted from being an affected aesthetic pose involving stillness and retreat to become more of an involuntary response to the exhausting hyperactivity and excessive production characterising contemporary life. Echoing Georg Simmel's early-twentieth-century theory of the blasé metropolitan attitude, in which individuals adopt a posture of indifference as a mechanism for coping with the psychological demands of modern city life, boredom in the age of speed derives from the pervasiveness and aggressiveness of creativity as an activity driving urban development and cultural-economic life in neoliberal (and neoliberalising) societies.[15] To put this another way, boredom is no longer a *precondition* for creativity, as per the Romantic conceit. Instead,

15 Georg Simmel, 'The Metropolis and Mental Life', in *The Blackwell City Reader*, eds. Gary Bridge and Sophie Watson (Oxford: Blackwell, 2002), 11-19. Originally published in 1903.

boredom has become an *outcome* of creativity, a result of overstimulation, overabundance, overexposure.

Fatigue

Thus, we can speak of the 'boredom of speed' – the cognitive, aesthetic fatigue that occurs in the face of rapid, constant newness. At one level, the rise of the global creative industries – including the transnational networks of communication and power in which they operate – have transformed the conditions under which creative practice occurs, eroding the autonomy of the artist, networking production, commodifying innovation, streamlining development, fast-tracking culture. Beijing's 798 Art Zone is a conspicuous example of this trend and one that, like many such initiatives worldwide, repurposes former factory buildings for use by artists and galleries. In this government-backed, policy-led, post-industrial creative complex, the presence of art serves to drive larger dynamics of gentrification and urban renewal, as well as to accelerate the commercial and aesthetic flows of the contemporary global art market.[16] The resulting slippage between creativity and profitability registers in a peculiar phenomenon that various commentators have observed: the proliferation of luxury sports cars, belonging to both artists and patrons, parked in the streets around 798. This conjunction of art and automotive bling speaks not only to the financialisation of creative practice, but also to the link between that practice and globalisation's culture of speed – quite literally embodied here by the excessively, impractically fast cars.

Yet, the 'boredom of speed' also manifests itself in other ways that are less complicit with the economic imperatives of neoliberal globalisation. Consider the example of vertical gardening with which I opened this essay. Like guerrilla gardening and urban farming more generally, vertical gardening is a community-oriented form of bio-urbanism that has emerged as a direct reaction against the speed of contemporary life and the flattening of creativity in urban environments. Vertical gardening's green insurgency does not emerge out of tranquillity. Rather, within a larger landscape of speed, vertical gardening seeks to generate conditions of tranquillity in which embodied, durational experiences such as boredom become newly possible through creative slow practice.

16 Jeroen de Kloet, 'Created in China and Pak Sheung Chuen's Tactics of the Mundane', *Social Semiotics* 20.4 (2010): 441-55.

1.4.3 The proliferation of luxury sports cars, belonging to both artists and patrons, in the streets around 798

Here, boredom is much more than an aesthetic strategy. It becomes a form of critique, however ephemeral and oblique, against globalisation's architecture of velocity. Yet, like the Californian slow-fast burger, the vertical garden in Hong Kong nonetheless remains ambivalently, vicariously connected to accelerated living. It also remains similarly reactive and, potentially, exclusive (it hinges on participants having free time). What matters is that the wall garden invites an escape *into* boredom rather than an escape *from* boredom, and in this respect connects to an emerging trend in urban counterculture aimed at recalibrating the speed-space of the contemporary city in order to revalue the human experience of time.

Note

I would like to thank Joyce Cheng for sharing her work on Paris Dada and boredom, and for introducing me to the field of interdisciplinary scholarship known as 'boredom studies'.

Bibliography

Barthes, Roland. 'The Death of the Author'. In *Image-Music-Text*, ed. and trans. Stephen Heath (New York: Hill and Wang, 1977).

Castells, Manuel. *The Rise of the Network Society* (Oxford: Blackwell, 2000).

Crary, Jonathan. *24/7: Late Capitalism and the Ends of Sleep* (New York: Verso, 2013).

de Kloet, Jeroen. 'Created in China and Pak Sheung Chuen's Tactics of the Mundane', *Social Semiotics* 20.4 (2010): 441-55.

Flew, Terry. *Global Creative Industries* (Cambridge: Polity, 2013).

Florida, Richard. *The Rise of the Creative Class* (New York: Basic Books, 2002).

—. *The New Urban Crisis: How Our Cities Are Increasing Inequality, Deepening Segregation, and Failing the Middle Class – and What We Can Do About It* (New York: Basic Books, 2017).

Honoré, Carl. *In Praise of Slow: Challenging the Cult of Speed* (New York: Harper Collins, 2004).

Lees, Loretta, Hyun Bang Shin, and Ernesto Lopez-Morales. *Planetary Gentrification* (Cambridge: Polity, 2016).

Lindner, Christoph and Miriam Meissner. 'Slow Art in the Creative City: Amsterdam, Street Photography, and Urban Renewal', *Space and Culture* 18.1 (2015): 4-24.

Perkins, David N. *The Eureka Effect: The Art and Logic of Breakthrough Thinking* (New York: W.W. Norton, 2001).

Simmel, Georg. 'The Metropolis and Mental Life'. In *The Blackwell City Reader*, eds. Gary Bridge and Sophie Watson (Oxford: Blackwell, 2002), 11-19.

Tomlinson, John. *The Culture of Speed: The Coming of Immediacy* (London: Sage, 2007).

Wordsworth, William. 'The Preface to Lyrical Ballads'. In *Lyrical Ballads and Other Poems* (Ware: Wordsworth Editions, 2003), 5-25.

Zukin, Sharon. *Naked City: The Death and Life of Authentic Urban Places* (New York: Oxford University Press, 2011).

About the Author

Christoph Lindner is Professor of Urban Studies and Dean of the Bartlett Faculty of the Built Environment at University College London. His recent books include *Imagining New York City* (Oxford University Press, 2015) as well as the edited volumes *Deconstructing the High Line* (Rutgers University Press, 2017), *Global Garbage* (Routledge, 2016), *Cities Interrupted* (Bloomsbury, 2016), and *Inert Cities* (I.B. Tauris, 2015).

1.5 Evaporating Ennui

Water Calligraphy in Beijing

Laura Vermeeren

Abstract
Early in the morning, groups of retired or middle-aged men and women flock to the public parks to cover the floors with calligraphy in all shapes and sizes using water instead of ink. This particularly public type of calligraphy started after the Reform and Opening-up policy at the beginning of the 1980s and spread from the capital of Beijing to public parks all over China. It is called water calligraphy, or ground writing in Chinese (*dishu* 地书) and takes up a significant part of the everyday lives of a growing number of people in China. This chapter offers an ethnographic account of the distinctive spatial, ephemeral, and social characteristics of water calligraphy.

Keywords: everyday life, vernacular creativity, calligraphy, public space

'This is just for fun! You have to do something when you get old, right?' A 79-year-old man told me in a public park in Beijing while he had just covered twenty metres of the park's surface with running script-style calligraphy written in water. His statement struck me with its common-sense simplicity. As I soon found out, he was not the only one holding this view. It was one of the most common responses I received from water calligraphers about their drive to practice water calligraphy in the park, where the majority of them spends almost every morning until noon writing on the floors.

A feeling of curiosity and astonishment drove the fieldwork I conducted on this practice: the first time I went to the public parks *Taoranting* and

Kloet, Jeroen de, Chow Yiu Fai, and Lena Scheen (eds), *Boredom,* Shanzhai, *and Digitisation in the Time of Creative China*. Taylor & Francis Group, 2019
DOI: 10.5117/9789462984745_CH1.5

Ditan in Beijing in the winter of 2015 and spring of 2016, I stumbled upon a group of older men, doing *something* with what looked like oddly shaped broomsticks in their hands. On closer inspection, I saw that they were writing with these tools, which turned out to be home-made brushes. The sight of a group of water calligraphers at work in the park is fascinating: seemingly effortless movements of the body and brush create beautiful characters glistening in the morning sun, vanishing again within minutes. In doing so, the empty tiles transform into slates of paper, while park visitors and calligraphers discuss the slowly evaporating brush strokes, the calligraphic styles and their content. The park transforms into a temporary place of literacy before your eyes. Further observing the other park activities in the morning, I felt like witnessing a celebration of some kind, not only of social life but also a celebration of the body. As early as seven o'clock, groups of retired or middle-aged men and women flock in to the park to dance, work out, talk, sing, laugh, and meditate. You will find many gatherings of singers, dancers, walkers, kite-flyers, chess-players, and *taijiquan* practitioners all over the space of the park. Most of them are retired, but among them you will find young parents with their toddlers as well. Among them are those ground-writers, men and women using hand-cut brushes made from sponges and broomsticks or umbrella sticks to letter the floors of the public park with calligraphy in all shapes and sizes using water instead of ink.

This research on water calligraphy is part of my dissertation on various practices of calligraphy and their societal significance in contemporary China. I had set out to, in an ethnographic mode, answer the questions that puzzled me about the water calligraphy phenomenon. Firstly about the people I found there: Who are they? Why are they here? Why do I see mostly older men? Then the practice itself: What are the constituent parts of water calligraphy, what seem to be the purposes and where are the practices of water calligraphy located? What national, political, and personal narratives underpin the construction of this new kind of calligraphy?

In what follows, I will offer a brief account of what these water calligraphers create, and what that, in turn, might create for the public space of the park. These accounts should be taken as scrapbook notes from the field, and its purpose is to offer a small peek into the doings of water calligraphers of the public park as I encountered them in the mornings of these months of 2015 and 2016 in *Ditan* and *Taoranting* Park just outside the second ring road in Beijing and *Taoranting* Park in the south-western corner within the second ring of Beijing.

Context

Both parks are frequented by a loosely formed group of around fifteen water calligraphers. The larger part of this group is male, and all of them are retired, ranging from the ages 55 to 81. Most of them come every morning for a couple of hours, some of them at least a few times a week.

This particularly public type of calligraphy started earlier than that, after the Reform and Opening-up policy at the beginning of the 1980s. It rapidly spread from the capital of Beijing to public parks all over China and groups of water calligraphers can be found in almost every public park in every Chinese city. It is called water calligraphy, or ground writing in Chinese (*dishu*, 地书). Chastanet has estimated that probably several millions of people are now writing water calligraphy in the park.[1] Xue Fengli, the Vice-President of the Water Calligraphy Association of Beijing in *Taoranting* Park (*Taoranting Dishu Xiehui*, 陶然亭地书协会) estimates that there are about ten thousand water calligraphers registered at various local water calligraphy associations nationwide. These associations convene annually for the water calligraphy competition in *Taoranting* Park in October in Beijing, with pre-selections in 26 different cities nationwide. Then there are many more water calligraphers that are not part of any organisation, which makes an accurate estimation difficult.

At first glance, we seem to be looking at what can be tentatively described as a curious hybrid between hobby-like physical exercise and a kind of calligraphic practice. Many of the people I spoke to have explained that seeing friends, practicing a skill, and moving the body is a good way to escape the ennui of retired life. I have heard often that it 'is boring to stay at home', and being outside and active is preferred. The water calligraphers are mostly retired. The relatively young retirement age in China, 50-55 for women and 55-60 for men attribute to the free time the 'elderly' have on their hands. With an average life expectancy of 75 years, there is, roughly speaking, twenty years of unemployed life to enjoy. Chen argues in her article on public dancing in Beijing that public exercise became popular as an unexpected result of the One Child Policy instituted in 1978.[2] As ageing Chinese now often have only one child, there are less children and grandchildren to look after, and more free time to spent. Water calligraphy seems to underscore the notion that creativity holds the promise of an escape out of boredom:

1 Francois Chastanet, *Dishu: ground calligraphy in China* (Stockholm: Document Press, 2013).

2 Tong Chen, *Square Dancing in the Streets, Xuanhua, China*. PhD dissertation (Columbus, OH: The Ohio State University, 2013), 39.

elderly people who would otherwise be at home, or possibly be bored, are now clustering together in the park to create something together. At the same time, boredom may also be an important constitutive condition for creativity: would these people have fabricated their own brushes and have spent morning after morning creating evaporative calligraphic characters if there was not a surplus of time? Supplementing to this boredom thesis may be habit, in the sense that the every day lives of these retired Chinese have been structured through a collectivist Maoist narrative, thus their preference for collective rather than solitary recreation.

Content

It is noteworthy that the public park is the space of choice for water calligraphers. Life inside the gates of the public park and outside on the streets are strikingly different, and spatial and social rules appear to be rewritten according to a different set of desires and needs. The water calligraphers will occupy ten to a hundred square metres of park surface depending on the size of their characters. They will chat with each other when they remoisten their brush and discuss each other's work and talk to passers-by about their written characters. The characters are less of a physical boundary to the space they occupy than one would expect – dancers and park strollers freely walk and dance over them, and the calligraphers do not seem to mind. The public park with its free entrance – retirees do not need to pay the small admission fee of two *yuan* – seems to suit the needs of the water calligraphers particularly well. There are many vast spaces with big grey tiles to write on, there is no shortage of attention and admiring looks of passers-by, and because they go to the same spot every day, it becomes a convenient meeting point.

The water calligraphers write in all calligraphic styles, but running script (*xingshu,* 行书) and grass script (*caoshu,* 草书) seem to be favoured because the flowing style in which it is written resonates with the flow of the water. Although ink is exchanged for water and expensive brushes and paper for trash-brushes and tiles, I observed that the stylistic rules of Chinese calligraphy are seldom negotiated. They write in traditional characters and most of the writers stick to writing in one style. Although in the Chinese mainland simplified characters are in use since the 1950s, in calligraphy discourse one writes in traditional characters. This shows that the people in the park are not just writing down characters – the deliberate use of traditional characters places them in the category of calligraphers. The content of their writing is congruent to 'normal' calligraphers' subjects of choice. Maoist poetry, Tang poetry and Chinese proverbs (chengyu) are favourite subjects.

The ephemeral nature of water calligraphy adds to its lure. It contradicts traditional calligraphic practice, which has as one of its main qualities that it resists time. Chinese calligraphy is a time-honoured traditional art; using, copying, imitating, and interpreting characters continuously for over three thousand years. Using bronze, stone, ink, and paper, Chinese calligraphy has managed to defy the 'disappearance' that is taking place with the characters written down in water. The main purpose of calligraphy has been communication, which quite obviously favours an endurable method of writing. That this type of calligraphy is ephemeral, one water calligrapher commented, is something new and not in line with how he, and in general, people would treat written characters in the past. He commented:

> Writing on the floor with water and sponge, ha. In Chinese tradition we have an old saying: 'Treasure the paper with characters on it'. You cannot just casually use a piece of paper that already had characters on it. Anything with characters on it is regarded 'old culture'. That's the way we thought before. You wouldn't write on the floor in case people would step on it.

Often people have commented on the sociable and classless quality of water calligraphy. One interviewee noted:

> Here we are all equal. Some of us are from the countryside, some of us used to be professors. Here we are each other's classmate and teacher at the same time.

Another calligrapher has been going to the park to write for two years. On strolling in the park, she was struck by the beauty of the characters and wanted to learn to do so as well:

> I did not finish high school, but now I can recite the Lanting Xu.[3] Whenever I get the chance, I go. First, they taught me how to write, and now we are all learning from each other. I feel so proud when people are passing by and compliment me on my writing.

3 *Lanting Xu* refers to 兰亭集序, Lantingjixu, and is regarded an exemplary calligraphic work written by the famous calligrapher, often referred to as the Sage of Calligraphy, Wang Xizhi (303-361). The piece, literally 'Preface to the Poems Collected from the Orchid Pavilion' is commonly perceived as a perfect example of calligraphy.

1.5.1 The character *guo*國, written in Ditan Park, Beijing

On several occasions the economic reasons for practicing in the park were stressed. One interviewee mentioned that there is really no need for ink, paper or even 'expensive lessons':

> Just buy a calligraphy book of Wang Xizhi to copy from for twenty yuan, and that is all you need. As long as you practice and persist, everybody can do it!

I think that exactly this – practice and persist – is what makes this practice significant for not only the study on (newly emerging) calligraphic

1.5.2 The character *yan*燕, written in Ditan Park, Beijing

practices, but also for further investigations on the productive affordances of boredom. While boredom seems to provide the time and space for this type of leisurely activity, the content has been provided by the normative institution of calligraphy. Water calligraphy does not resist violently what they have grown out of but creates for itself the space to celebrate a calligraphy that can be practiced by anybody who does not want to be bored at home, and has the time, space, and persistence to imagine a new kind of calligraphy.

I will conclude with a poem written by water calligrapher Huang Songbai who wrote a poem in the Taoran water calligraphy Newspaper of January 2016:

We wrote characters on oracle bones – the immortal oracle script
We wrote characters on magnificent bronzes
We wrote on glorious paper
Today we boldly dip our brushes in clear water for our leisure
reside with poetic exuberance on the earth

1.5.3 Learning how to write *de*德, written in Ditan Park, Beijing

1.5.4 Writing in the winter makes the characters freeze

1.5.5 A retiree enjoying himself in Taoranting Park, Beijing

1.5.6 Writing running script in Ditan Park, Beijing

1.5.7 Water Calligraphy movie

https://www.youtube.com/watch?time_continue=3&v=E_8lj_nhRio

This research is supported by a consolidator grant from the European Research Council (ERC-2013-CoG 616882-ChinaCreative).

Bibliography

Chastanet, Francois. *Dishu: ground calligraphy in China* (Stockholm: Document Press, 2013).

Chen, Tong. *Square Dancing in the Streets, Xuanhua, China*. PhD dissertation (Columbus, OH: The Ohio State University, 2013).

About the Author

Laura Vermeeren is a Ph.D. candidate at the University of Amsterdam. Her research is concerned with how Chinese calligraphy as a living daily practice is entangled with technologies of the self, government and society in contemporary China. Previously she completed a Bachelor and Master in Chinese studies at Leiden University. She published peer-reviewed article *Chinese Calligraphy in the Digital Realm: Aesthetic Perfection and Remediation of the Authentic* (2017, Concentric), and Book Chapter "*We are Not Like the Calligraphers of Ancient Times" – A Study of Young Calligraphy Practitioners in Contemporary China* (2019, forthcoming with Professor de Kloet).

1.6 Male Cinderella on the Small Screen in mainland China

Li Hao

Abstract

The male Cinderella phenomenon has emerged on Chinese TV in recent years and appears to reverse the hegemonic representation of gender power relationships on the small screen in mainland China. Using Gramsci's framework of hegemony and alternative hegemony, this study focuses on the gender landscape expressed in TV dramas and explores the social meanings embedded by connecting the relationships between gender, the TV industry, and the social context in mainland China. Tracing the gender development in China, it is clear that the gender discourse is influenced by tradition, the market, and the state. Therefore, studying the male Cinderella phenomenon is a chance to see how commercial logic interacts with these three aspects of gender discourse in China.

Keywords: male Cinderella, Chinese TV drama, gender discourse in China, gender hegemony

The moment when I came up with the idea for the male Cinderella phenomenon on the small screen was inspired by the famous actor Huang Lei in mainland China, who dumped his former screen images as an intellectual to take on the role of a 'stay-home husband' in the *Marriage Battle* (婚姻保卫战) produced in 2010. His wife in the story is a powerful and successful businesswoman, running a big enterprise in Beijing. The couple's relationship and family style became a hot topic that year when the TV series seemed to challenge the traditional gender relationship in love and marriage in the Chinese society, which always emphasises the supreme power of maleness. Confucianism tends to promote the notion

Kloet, Jeroen de, Chow Yiu Fai, and Lena Scheen (eds), *Boredom,* Shanzhai, *and Digitisation in the Time of Creative China.* Taylor & Francis Group, 2019
DOI: 10.5117/9789462984745_CH1.6

that 'man works outside while woman stays inside'. This is the first time I find something like subversion on the small screen in my memory and to me, its appearance can be regarded as a counter-hegemonic existence to the sensational and long-lasting Cinderella stories on the TV screen before. I tentatively term it 'the male Cinderella' in China, which is the topic of my research project.

Cinderella changing gender

The Cinderella story is one of the most popular drama genres not only in China, but around the world. Among all the Cinderella myths, the Grimm brothers' version is the one that has most fans in the world. It is believed to be one of the most popular reading materials for girls who nurture dreams of becoming a princess. Moreover, it almost becomes a faith that beautiful and kind girls can be lucky enough one day to meet a prince.[1] After all, good girls can always make their dreams come true. Cinderella is 'usually portrayed with positive emotions. She is kind, helpful, sympathetic, and loving, and by implication, pure and good'.[2] When a story features a comparatively weak girl meeting a rich and powerful boy, it falls within this Cinderella romance frame.[3]

The booming of contemporary Cinderella TV dramas in mainland China began with the Taiwanese idol drama *The Meteor Garden*（流星花园）in the early 2000s, when the metrosexual boy band 'F4' became a hit. The drama centres around a romance between an ordinary girl and a handsome boy from a super wealthy family. Although the social class distinction always obstructs their romance, the boy's unrequited love finally wins the girl's heart.

The success of the idol drama has inspired the TV drama industry in mainland China to create similar stories. From the early 2000s, TV audiences switching on are guaranteed to find Cinderella stories. To name a few, *Boss and me*（杉杉来了）, *Cruel romance*（锦绣缘华丽冒险）, *Love & life & lie*（遇见爱情的利先生）and *Pretty Li Huizhen*（漂亮的李慧珍）, all belong to this category. Initially, these Cinderella stories were hugely popular, especially among women viewers who could imagine

1 Alleen Pace Nilsen, 'Women in Children's Literature', *College English* 32.8 (1971): 918-26.

2 Karol Kelly, 'A Modern Cinderella', *Journal of American Culture* 17.1 (1994): 87-92.

3 Elisabeth Rose Gruner, 'Saving Cinderella: History and Story in Ashpet and Ever After', *Children's Literature* 31.1 (2003): 142-54.

sharing the same life experience with the Cinderellas in the TV series, which can be understood as 'empathy'.[4] However, with the large number of Cinderellas on the small screen, inevitably negative comments and complaints began to appear on the internet, from the traditional discussion groups and bulletin boards to *Sina Weibo* platforms and WeChat public accounts subscribed to by millions of users. People complained that they were bored of the Cinderella stories especially when the plots and lines were so predictable. Moreover, now that video websites with the bullet screen function, like *Bilibili* and *AC Fun*, have become popular among the younger generation, people get easily bored of the traditional Cinderella TV dramas. Meanwhile, viewers seem to be attracted to other formulas of romance, such as *The overbearing boss falls in love with me* (霸道总裁爱上我),[5] where the typical Cinderalla division of romantic labour is somewhat twisted. Here, I observe the beginning of change, from the conventional Cinderella to a mocking Cinderella or what I call a male Cinderella version. The surge in popularity of the latter formula affirms that the Chinese audience are weary of shows whose storylines are predictable and do not engage or challenge viewers.

The advent of the male Cinderella story, I suggest, is a response to the incessant repetition of Cinderella TV dramas and represent a new type of love relationship in contemporary China. The male Cinderella can be understood as a beta male, who is 'mojo-free, easy going, soft spoken and happy following'.[6] Moreover, these characters are generally considerate, loving, and self-sacrificing. The trend for the male Cinderella TV drama is promoted by the company New Classics Media, which says it produces the TV series from a female perspective. The male Cinderella TV dramas are usually produced by the same crew with the same director, producer, and scriptwriter (Yangzhe, 2016).[7] Cases like *Honeybee Man*（我爱男闺蜜）, *May-December Love*（大丈夫）, and *May-December Love II*（小丈夫）, all achieved success and high audience ratings during the periods when they were first broadcast. In order to understand the popularity of the male Cinderella TV drama, three semi-structured interviews were conducted in Beijing with the

4 Glen Creeber, '"Taking Our Personal Lives Seriously": Intimacy, Continuity and Memory in the Television Drama Serial', *Media, Culture & Society* 23.4 (2001): 439-55.

5 'The overbearing boss falls in love with me' is a popular phrase used on the Chinese internet to describe the modern Cinderella genre in which the prince charming becomes a super rich boss and the Cinderella is his employee. In the story, the pretty and pure employee is persistently pursued by her boss.

6 Jennie Yabroff, 'Betas Rule', (2007): 64-5.

7 Yang, Zhe. Personal Interview. June 9, 2016.

producer Huang Lan, the director Yao Xiaofeng, and the screenplay writer duo Li Xiao and Yu Miao, all employed by New Classics Media.

Unexpected Success

The male Cinderella TV dramas usually tell the same story: the female protagonists occupy a higher social status and control more resources than the male protagonists. Sometimes, the women are older than the men, breaking the rules of the ideal match, i.e. that a man should be older than a woman in a relationship in Chinese society. Nevertheless, they fall in love.

The success of the male Cinderella TV drama came as a surprise to the producers. According to the scriptwriter Li Xiao,

> We have never expected the success of the elder sister and younger brother's love (姐弟恋) in the *May-December Love*. We added this storyline in order to attract a younger audience, because other storylines in the TV drama are quite traditional. But later, I found that people care about other characters' love stories in my drama. According to the comments on *Weibo*, they focus on the development of untraditional love instead (Li Xiao, 2016).[8]

According to producer Huang Lan,

> The success of *May-December Love* inspired *May-December Love II*. Although we use the same cast, it is another story. And you can see the differences in the relationships. In *May-December Love*, we leave the audience with an open ending, while in *May-December II*, the love between the 'younger brother' and 'elder sister' ends happily. We believe our audience may be more able to accept this as time passes (Huang Lan, 2016).[9]

Fantasy or Reality

Considering the existence of Cinderella stories and the reality of male dominance in Chinese society, the appearance of the male Cinderella seems to be a challenge to the social convention, provoking questions about women's

8 Li, Xiao. Personal interview. November 9, 2016.
9 Huang, Lan. Personal interview. Novemebr 1, 2016.

empowerment. Moreover, almost all the TV dramas end with hope and happiness, when the couple conquers social distance. Even when women are more powerful than men, the women can still ultimately realise their dreams. However, when asked about whether these dramas truly reflect the social reality or simply create a fantasy for the audience, the interviewees give a range of answers.

From the scriptwriters' perspectives, they believe their stories are based on the social reality because 'you may actually find more examples in the real life when you see a couple of powerful woman and soft man. And the soft male images seem to become very popular in recent years' (Li Xiao, 2016).[10] In addition to social reality, they also refer to Chinese traditions as background to understanding the male Cinderella story: 'Also people can find similar examples in the traditional Chinese literature like "a gifted scholar and a beautiful lady" (才子佳人). In all these stories women have a higher social status. Therefore, in my eyes, the Chinese women are always powerful' (Yu Miao, 2016).[11]

On the other hand, the creators of these male Cinderella dramas also highlight the fantasy dimension. 'However, I admit a story like *Honeybee Man* is truly a fantasy, because you cannot imagine a man from the bottom finally winning the love of such a beautiful and capable woman' (Yu Miao, 2016).[12] In the eyes of the director, 'I do these TV drams because I want to make the audience believe in love in the contemporary society, when upward mobility is low and economic inequality is solidified, which makes the young generation feel pessimistic about love. I still believe in the gendered division of labour in the society, in that way, I did not make the story too deviant from the convention, so the *May-December Love* had an open ending' (Yao Xiaofeng, 2016).[13]

As my research is still on-going, I would like to conclude tentatively that the emergence of the male Cinderella TV dramas was not planned, but rather they were a reaction against being bored by traditional Cinderella stories and originated in a longing by TV professionals to create something non-traditional. On the other hand, these male Cinderella TV dramas often occupy the middle ground of gender stereotypes, somewhere between tradition and subversion, in order to ensure they are accepted by the mainstream audience.

10 Li, Xiao. Personal interview. November 9, 2016

11 Yu, Miao. Personal interview. November 9, 2016.

12 Ibid.

13 Yao, Xiaofeng. Personal interview. October 27, 2016.

Bibliography

Creeber, Glen. '"Taking Our Personal Lives Seriously": Intimacy, Continuity and Memory in the Television Drama Serial', *Media, Culture & Society* 23.4 (2001): 439-55.

Gruner, Elisabeth Rose. 'Saving Cinderella: History and Story in Ashpet and Ever After', *Children's Literature* 31.1 (2003): 142-54.

Huang, Lan. Personal interview, Nov 1, 2016.

Kelley, Karol. 'A Modern Cinderella', *Journal of American Culture* 17.1 (1994): 87-92.

Li, Xiao. Personal interview, Nov 9, 2016.

Nilsen, Alleen Pace. 'Women in Children's Literature', *College English* 32.8 (1971): 918-26.

Yabroff, Jennie. 'Betas Rule', (2007): 64-5.

Yang, Zhe. Personal Interview. June 9, 2016.

Yao, Miao. Personal interview, Nov 9, 2016.

Yao, Xiaofeng. Personal interview, Oct 27, 2016.

Links

http://v.youku.com/v_show/id_XNjgwOTk1NDE2.html?from=s1.8-1-1.2&spm=a2hok.8191407.0.0

http://v.youku.com/v_show/id_XMTU2MzQ2NjI5Mg==.html?spm=a2hoj.8191423.item_XMTU2MzQ2NjI5Mg==.A

http://v.youku.com/v_show/id_XNjk0NTA0MzI0.html?from=s1.8-1-1.2&spm=a2hok.8191407.0.0

About the Author

Li Hao received her PhD degree in communication from the Chinese University of Hong Kong and her current research fields include Chinese entertainment industry, sports, and e-dating.

1.7 'Performativity' in the Age of Banality

Lo Yin Shan, Kingsley Ng, and Kung Chi Shing

Interviewer: Lo Yin Shan
KN: Kingsley Ng (HK artist/Assistant Professor at the Academy of Visual Arts of Hong Kong Baptist University)
KCS: Kung Chi Shing (HK composer and music activist/ curator of West Kowloon Cultural District)

Q1: Whether a life (like you two as artists), fully occupied with creative activities and critical thinking, normally has less chance to 'encounter' or no place for 'boredom' (a term in a more general understanding)?
KN: I think this is an overly romanticised notion of art making – in every art project, it is inevitable that there exist various levels of tedious works beyond the creative activities, like writing proposals, budgeting, logistic, resource and production management, health and safety and risk management, insurance, auditing, outreach and education programme, project evaluation, media, PR and interviews, so on and so forth.
KCS: My life as an artist is so full of uncertainty, unknowns, risks and challenges that I don't really have time for boredom.

Q2: Or as an artist who is always living with curiosity and passion, so he won't perceive so-called boredom as other people do? Could you give an example of any 'creation of boredom' [(non-) materialistic, or eventful, or phenomenal], which has special meaning to you?
KN: I am not so sure about the above statement either. I do occasionally get bored by some of things mentioned in the above question. But very often, even the most insignificant logistical measures, such as the lighting or air quality of the entrance hallway, to the attitude of the docents of the artwork, may have an effect on the quality or consistency or conceptual reading of

Kloet, Jeroen de, Chow Yiu Fai, and Lena Scheen (eds), *Boredom,* Shanzhai, *and Digitisation in the Time of Creative China.* Taylor & Francis Group, 2019
DOI: 10.5117/9789462984745_CH1.7

1.7.1 Replicas of White House and Gate of Tiananmen in different towns of China

the work. Sometimes, these tedious works become part of a holistic artistic orchestration, and may not be too boring after all.

KCS: Boredom has to do with one's patience, tolerance, and perception of time.

Boredom doesn't motivate me to create art, and I doubt that it can be a motivation for most artists, but I do think 'boredom' can discourage or stop the audience's acceptance of art that is unconventional or stretched over a long duration (most people can stand one minute of 'strange' sound/ images, but if it goes on for a long stretch of time without making sense to the audience, he or she may find it boring and reject it.)

I think John Cage's works challenged people's conventional value system, such as the meaning of art, perception of time, the notion of boredom, the concept of beauty, etc.

1.7.2 A Chinese impersonation of Obama for an online TV drama

Q3: In the conference, Kingsley presented the book BOREDOM: The Literary History of a State of Mind *(1994), the author Patricia Meyer Spacks points out that the discontent of 'nothing is happening' motivates the literary imagination. For you, this kind of 'nothing is happening' could be regarded as a way of self-introspection or a space for meditation?*

KN: For audiences moving through galleries in a hurry, my installation work *Moon.gate* 閒 (2011) can be easily overlooked and be perceived as 'nothing is happening'. The work is always set in an empty room with architectural openings, and the audience can contemplate a projected simulation of light through the window frames. The shadows show a local tree, which sways in the hiss of the wind. While the sound actually comes from a tuning radio, the tree stays still when broadcast content becomes audible. In a blurry threshold between the simulated and the real, the idyllic is intertwined with current affairs. In the whispers of the light lies a fundamental question: what is the state of mind required for listening to the world we live in today?

KCS: Creating something (let's say art) is the opposite of 'nothing is happening'. The process of creating can have a concrete result (for example a piece of music or a film) and the artist may benefit from that process intellectually or spiritually. But I am not sure we are creating art simply because 'nothing is happening'.

Q4: And Kingsley used the character 悶,[1] *to refer a more illustrative mental status, do you think that our state of mind might be 'restrained' or differentiated according to the language we use e.g. in Chinese, English, or French thinking or when it is channelled in a form of visual or musical language?*
KN: I wouldn't say it is 'restrained', but certainly each language or translation carries complex nuances.
KCS: We are all subjected to cultural conditioning somehow, and language is closely related to its culture.

However, I think 'boredom' is a relatively 'young' phenomenon, very much related to a high-tech, modern society. Technology speeds things up exponentially, we can access any information with a click on the computer or smart phone. We can no longer sit still.

Q5: In the conference, Kung's presentation on 'How our notion of "boredom" hinders the development of new music' used minimalism, experimentalism, and reductionism in contemporary music as examples to show that experimental music is aimed at challenging rather than entertaining, i.e. to 'agitate/provoke' rather than to please the audience. So, 1. The seemingly passive concept 'boring' in fact has a more active 'function'/paradox, e.g. as a catalyst waiting for action? 2. Do you think this function/definition of 'boredom' in experimental music and art has varied from the 60s to the 80s, and from the 90s to 2016?
KN: I will let Kung answer this one.
KCS: Yes, I think the notion of boredom has changed over the years, I think it is related to the 'speed' at which we are living. We live at a much faster pace than people in the 60s, 70s, or even five years ago.

Q6: As a resistance against preoccupation with 'nothingness', we have continuity of 'happening', with all kinds of overloaded infotainment under the umbrella of the 'creative industry'. 'Art and culture' seems ubiquitous and, to a certain extent, has become a sort of 'decoration' for shopping malls, that's one aspect of 'performativity in the age of banality'. The banality is, as Kung said in an interview, that everyone can be a musician or an artist but quantity is not equal to quality. And 'performativity' is, on the one hand, designed to catch 'attention' among banality, or, on the other extreme, staging 'performance' is a kind of artivism. Do you think this is a paradox of our boring reality?
KN: I do share some of your observations, in answering to a world in which we are bombarded with spectacles, the deliberation of works such as *moon.*

1 Chinese character for boredom, figuratively 'a heart trapped inside a door'.

1.7.3 A 'mockup' bank and 'fake' police as illegal bussiness

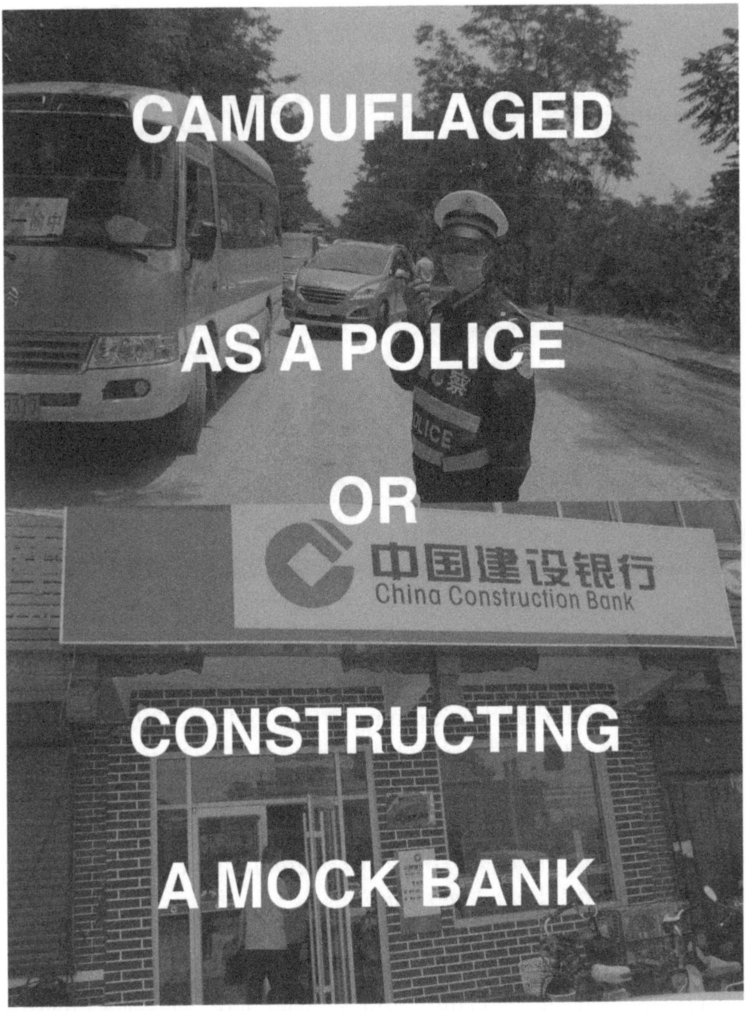

gate, is to present an empty space, where matters are not made to be seen, but made to be discovered. On the contrary, in contemporary times, instead of a resistance against preoccupation with nothingness, nothingness is becoming something that has to be acquired consciously through unconventional means.

KCS: I don't think art is a medicine to cure 'boredom' in modern society. I am a traditionalist, art is something pure and essential, just like love. It doesn't matter how the external world has changed, art's essence is the same. Art helps us to stay human.

1.7.4 The Chinese government once tried to blacklist the 'shanzhai' business

About the Authors

Lo Yin Shan graduated with a diploma in design from the Hong Kong Polytechnic University, and is currently a feature reporter/editor and an amateur artist/photographer. She lived in Beijing from 2008 to 2017 as a 'teleporter'/columnist between two cities. Author of *Driving Lantau: A Whisper of an Island*; *Ten Notes on Northern Drifting*; editor of *The Box Book: Beauty and the Beast of Hong Kong Culture.*

Kingsley Ng is an interdisciplinary artist and designer with a focus on conceptual, site-specific, and community-engaging projects. His creative practices are driven not by a self-indulgent romance of art, but a belief that art can be socially relevant. Ng holds a Bachelor of Fine Arts degree in New Media from Ryerson University, Canada; a Master of Science degree in Sustainable Design from The University of Edinburgh, Scotland; and Post-Diplôme avec les félicitations du jury à l'unanimité (Master of Fine Arts degree equivalent) from Le Fresnoy – Studio National des Arts Contemporains, France. He is currently Assistant Professor at the Academy of Visual Arts of Hong Kong Baptist University.

Kung Chi Shing, born and raised in Hong Kong, is a composer, performer, and music activist. Kung studied classical music and composition in the United States with Allen Trubitt and George Crumb. He is known particularly for his work in the music and performance group 'The Box', which he founded with Peter Suart in 1987, and more recently as founder and curator of the 'Street Music Series' of free public concerts in Hong Kong. In his compositions and performances, Kung focuses on experimentation with different formats, including pop, classical, and improvisational music, and theatre art.

Section 2

Shanzhai

Essays

2.1 New Productive Culture

Shanzhai or Second Degree of Creation?

Anthony Fung and Yiyi Yin

Abstract

This chapter describes cultural production by the new generation in China and conceptualises it as 'second-degree reproduction' based on what we call '*shanzhai*', referring to an imitation or copycat. In the second-degree reproduction of cultural products as foreign popular culture in the digital space, Chinese youth are able to create an alternative discourse based on the existing controlling narratives in society to challenge the dominant mainstream. While this alternative is seen as '*shanzhai*', it remains marginal in the eyes of the authorities and thus is free to stay intact, reproduce, and regenerate. Through analysis of concrete examples on social media, this paper explores the nature of such second-degree cultural production.

Keywords: *shanzhai*, second-degree reproduction, youth culture, globalisation

Introduction

The discussion of transcultural reproduction is never new in the era of globalisation. As exotic cultural products flow around the world especially through mass media and digital channels, the youth generation who grew up during a period that witnessed the acceleration of globalisation has formed a bicultural identity in which both global and local culture are

Kloet, Jeroen de, Chow Yiu Fai, and Lena Scheen (eds), *Boredom,* Shanzhai, *and Digitisation in the Time of Creative China*. Taylor & Francis Group, 2019
DOI: 10.5117/9789462984745_CH2.1

mingled.[1] Nevertheless, it would be dangerous to assert that regional markets have all opened to foreign cultural products to the same extent. In specific countries like China, rigid policies and marketing strategies remain powerful in terms of isolating the local market and resisting against globalisation. Admittedly, the state has gradually opened its market to foreign products since its accession to the World Trade Organization (WTO) in 2001. However, most of the opening policies are short-term and with obvious restrictions. By promulgating a series of policies that is deemed to be anti-globalisation, the state is still holding strong control of foreign cultural production imported into the local market. Strong ideological censorship is held in industries such as film, TV, animation, newspaper, and even the internet, to halt the import of 'unhealthy' or politically incorrect foreign productions.

To a certain extent, the progress of globalisation seems to be controlled and restricted in China. Foreign culture and products are not, of course, completely rejected by the country. Nevertheless, the policies do prevent domestic consumers from freely choosing cultural products on a global scale. Under such circumstances, the phenomenon of *shanzhai* or 'second-degree reproduction' as it will be referred to here, emerged as a new form of globalisation in China. Instead of the localisation of traditional imported media products or capital such as Hollywood films or McDonalds, second-degree reproductions appears as a self-generated and domestic youth cultural practice that copies, clones, and imitates global culture.

The concept of 'second degree reproduction', as it will be discussed later as the core concept in this chapter, has its similarities with the concept of *shanzhai*, but holds its own features at the same time. Literally referring to 'mountain villages with fences and houses'[2] in Chinese, *shanzhai* is broadly seen as a copycat culture that reflects the process of modernisation and the potential of grass-roots creativity in the country.[3] The word *shanzhai* is thus

1 Jeffrey Jensen Arnett, 'The Psychology of Globalization', *American Psychologist* 57.10 (2002): 774.

2 Lin-chuan Qiu, '*Shanzhai* Culture in the Era of the Internet', *21 Century* 112 (2009): 122.

3 Qiu, '*Shanzai* Culture'; Jing Wu, '*Shanzhai* Economy and New Media Creativity – A Discussion of the Formation of Postmodernist Public Sphere', *China Media Report*, 2010, accessed August 28, 2016, http://www.chnmedia.com/wordpress/?p=2578; William Hennessey, 'Deconstructing *Shanzhai*-China's Copycat Counterculture: Catch Me If You Can', *Campbell L. Rev.* 34 (2011): 609; Lin Zhang and Anthony Fung, 'The Myth of "Shanzhai" Culture and the Paradox of Digital Democracy in China', *Inter-Asia Cultural Studies* 14.3 (2013): 401-16.

usually translated as 'parody',[4] referring to the entertaining and popular copy of a cultural product. Similarly, Gong and Yang define Chinese parody as an 'alternative locus of power' that has the potential to resist the authoritative discourse.[5] On the one hand, it means that the format of *shanzhai* products is usually informal, recreational, and even vulgar on some occasions. On the other hand, it implies a reproduced creativity embodied in the process of 'making fun of' or 'parodying' the original product. Nevertheless, the popularity of a *shanzhai* product largely relies on its close connection with the original product or brand. This is precisely the reason why *shanzhai* products in China always have similar names to the products they are copied from. Famous examples include 'abidas', copied from 'adidas', and the online Spring Festival Gala, which is obviously the 'twin' of CCTV's Spring Festival Gala show.

To a large extent, second-degree reproduction is a more radical cultural reproduction than *shanzhai*, because it actually attempts to emphasise a kind of autonomy and self-owned authority, rather than just becoming an entertaining and cheap appendant of the existing products. As the chapter will further discuss, the *shanzhai* culture in popular culture, particularly within the context of globalisation, has become a form of second-degree reproduction that multiplies global contents internally by producing and consuming *shanzhai* products in specific subcultural or non-official spaces. Instead of being officially imported and localised, foreign culture is redone and copied in local products that are produced and consumed domestically by youths, especially by fans of foreign popular culture. Depicting cases mainly from the reproductions of East Asia, especially Japanese and Korean popular culture in China, the chapter examines this cultural practice and its further implications, arguing that second-degree production must be studied as a new form of globalisation with both its countercultural potential for Chinese youth, as well as its challenging relationship with multiple powers, especially the cultural industries.

Consumption of Japanese Anime in China

Japanese popular culture, especially animations, are highly popular among Chinese youths who were born after the 1980s when *Astroboy* was introduced

4 Zhang and Fung, 'The Myth of "Shanzhai" Culture'.

5 Haomin Gong and Xin Yang, 'Digitized Parody: The Politics of *Egao* in Contemporary China', *China Information* 24.1 (2010): 4.

into the country.[6] Over a period of twenty years, when the importing policies were relatively relaxed, Japanese anime developed into one of the most influential subcultures embraced by large numbers of fans in China. Popular series such as *Touch, Sailor Moon, Pokemon*, and *Slam Dunk* were broadcast on TV for years. Pirate manga bookstores sprang up across the country. The emergence of the internet enhanced the popularity of Japanese animation and facilitated the formation of fan groups and related activities both online and offline. Anime clubs appeared to be popular among college students, along with the emergence of comic events, also organised by college students (Sonna, 2014).[7] For instance, Comic Dive, one of the largest indoor comic events in Beijing, was organised initially by animation clubs from Peking University and the Communication University of China in 2008.[8] Another comic event, Linjie, has also been hosted by the anime club of Renmin University in Beijing since 2005.[9] Today, hundreds of comic events and animation festivals are held in cities like Beijing, Shanghai, Chengdu, and Hangzhou each year, including those that have become crucial events in domestic cultural life. In 2015, the Chinese International Cartoon & Animation Festival, held in Hangzhou, attracted more than 1.3 million visitors with over fourteen billion RMB of sales.[10] Basically, Japanese anime and the culture behind it have played an important role in Chinese youth's cultural consumptions and practices.

Nevertheless, for political and economic reasons, the government has established a series of policies relating to importing Japanese animation in recent years. These policies can, on the one hand, be understood as a national strategy in the era of globalisation. To develop and protect the local cultural industry and the related 'soft power', the policy attempts to prevent 'cultural imperialism' in the market and to encourage domestic creativity. On the other hand, as is evident from the 'blacklist' of foreign products, a particular ideology might be politically sensitive to the authorities or the government. Consequently, in 2000, the administration restricted the amount of animation productions that could be imported for TV. In 2006,

6 Zhong-Afu Guan, 'Where is the Path? How Can Japanese Animation Step on Chinese Road?', *Insights*, April 29, 2015, accessed September 4, 2016, http://culture.ifeng.com/insight/special/anime/.

7 Sonna, '中国 ACG 圈的发展历史是怎样的，有哪些重要网站', Zhihu 2004. Accessed March 10, 2019. https://www.zhihu.com/question/22561605/answer/25534694.

8 See 'Comic Dive': http://baike.baidu.com/link?url=BZqn_EYTpWLypLQtn6cly9SzJfdcH-WPko6K_xTXeADlGXidRPBMHwNQNIIuSuymoA5Giox-7poHm8apfJB_wH_.

9 See 'Linjie': http://baike.baidu.com/subview/4780554/4780554.htm.

10 CICAF, 'Introduction of the Chinese International Comic & Animation Festival (CICAF 2015)', accessed August 28, 2016, http://www.cicaf.com/xwzx/content/2015-01/01/content_5576429.htm.

policies further forbade foreign animation from being aired on TV during the 'golden period', from five pm to nine pm.[11] More radically, in 2015, the government banned over 40 Japanese animations from TV. The listed animations, which are forbidden due their 'indecent, violent, criminal or horror' content, which 'may negatively influence Chinese audience cognitively', include very popular blockbusters such as *Death Note*, *Blood-C*, *Attack on Titan*, and *Assassin Classroom* (Guan, 2015; Administration of Culture of the PRC, 2015).[12] In past decades, the TV broadcasting of foreign animations, especially Japanese anime, has been strictly controlled and censored by the government, in the name of protecting domestic youth from 'poisoned content'.

For youths who grew up in the late 80s and 90s, this kind of ideological control policy directly threatens the daily cultural consumption of anime that they are so attached to. Under such circumstances, the consumption of Japanese animations and related products was turned into an underground practice that results in fans playing 'hide and seek' with the administration. In the earlier times, people had access to anime and manga by renting or purchasing pirate copies at illegal DVD stores. Later, fans were able to download video sources directly from fan-subgroups, who record the Japanese original video and translate it themselves. More recently, animation video sites like *ACFun* and *Bilibili* offer more stable and safer online spaces for animation fans to post and watch animations. To circumvent censorship, tricks such as changing the name of *Assassin Classroom* into *Grade 3, Class E* were played by fans. In doing so, fans are still able to enjoy limited animations, no longer from official channels, but in their so-called local space that is shaped by their own efforts to record, translate, and post content. Remaining subcultural and marginalised, such local spaces keep a relatively safe distance from the government's control and restrictions but are embraced by youths and fans as their main habitat in cultural life. In this habitat, youths have access to the foreign products they are fond of and are able to discuss them with others.

Initially, online communities such as *Baidu Tieba* and other theme forums were major digital spaces for youth to talk to each other about certain novels, animations, and other forms of popular culture. Nowadays, newer internet platforms, such as *ACFun* and *Bilibili*, have been established to

11 Guan, 'Where is the Path?'; Xinhuanet, 'New Rules by Administration of Broadcasting and Television: Foreign Animations are Forbidden to Air on TV during Golden Period', *Xinhuanet*, 2006, accessed September 4, 2016, http://news.xinhuanet.com/politics/2006-08/13/content_4954807.htm.
12 Guan, 'Where is the Path?'; Administration of the Culture of the PRC, 2015.

allow a higher level of participation than merely discussion. These local spaces, which function as platforms for fans to share their own thoughts and works, also encourage the emergence of certain forms of fan creation. Being motivated by strong affection, these creations started from so-called textual poaching, which refers to the rewriting or re-edition of the original text based on fans' own blueprint.[13] In their continued struggle against the harsh importing policies and the limitation of accessible Japanese anime, such 'textual poaching' is gradually developing into second-degree reproduction that is not only a consequence of affection, but also a consequence of the anti-globalisation policy. Such reproductions do not limit themselves to 'poaching' and 'rebuilding' subtexts based on the original animation or manga, but also create brand new products carrying the qualities and characteristics they love. Consequently, second-degree reproduction is more closely related to *shanzhai* culture and should be considered as a bottom-up and internal version of globalisation, an expression of grass-roots creativity rather than official culture.

Copycat Culture Revisited

The *shanzhai* culture in China can be traced back to its traditional meaning in imperial China, where it referred to mountain villages occupied by outlaw bandits who were evading the imperial court.[14] In the 1960s, *shanzhai* referred to the underground factories in Guangdong and Hong Kong that produce poor-quality and cheap commodities.[15] The later boom in *shanzhai* mobile phones in the early 21st century re-identified the term *shanzhai* as the representation of copycat culture in China, implying the cheap parody of expensive and branded products, especially electronics. A result of the developing economy in the country, 'where global name brand commodities have established their luxury status through profit margins that put the vast majority of the people below the threshold of consumption',[16] this kind of copycat practice and production emerged initially to produce status products at lower prices. Nevertheless, along with the huge popularity of those *shanzhai* products in the domestic market, especially in smaller cities

13 Henry Jenkins, *Textual Poachers: Television Fans & Participatory Culture* (New York: Routledge, 1992).

14 Qiu, '*Shanzhai* Culture'; Zhang and Fung, 'The Myth of "Shanzhai" Culture'.

15 Qiu, '*Shanzhai* Culture'.

16 Josephine Ho, '*ShanZhai*: Economic/Cultural Production through the Cracks of Globalization'. Plenary presentation (Hong Kong: ACS Crossroads: 2010 Cultural Studies Conference, 2010), 2.

and villages, *shanzhai* is increasingly understood not only as the production of low-quality imitations, but also as an icon for innovative copycat production and grass-roots rebellion whose unique form of creativity resists 'upper-class' ideology and officialdom.[17]

Although studies have argued that *shanzhai* has the potential to evoke alternative innovation in China, rather than simply copying existing products,[18] the cultural implications of the term need to be examined both socially and historically. Closely related to the Chinese traditional ethos of 'robbers as heroes', *shanzhai* is explained by some discourses as the continuity of folk creativity that counters the official Confucian culture in imperial China.[19] As William Hennessey, *shanzhai* emerges as a reaction to the dominant control in terms of both economy and ideology. Similar to humorous stories and popular fictions during the imperial period, contemporary *shanzhai* can also be seen as a counterculture that relies heavily on folk creativity and grass-roots intelligence to speak against the Chinese official bureaucratic culture and its value system. The *shanzhai* Spring Festival Gala, for instance, cultivates an alternative democracy that has embedded political resistance into affection, collective creativity, and self-branding.[20] By making a joke about and playing with dominant ideology and established 'high-brow' products, *shanzhai* production gives a voice to the marginalised and trivial spaces that are different from the mainstream and official discourse.

Global products play an important role in this process. The copycatting practice in *shanzhai* culture has a long tradition of imitating foreign products such as McDonalds, iPhones, and Nokia mobiles. On the other hand, as some discourses have suggested, *shanzhai* is at the same time a counterculture, copying luxury goods and turning them into cheap, low-quality products in opposition to the Western 'global innovation culture' and the capitalist system.[21] In the field of popular industry in China, where cultural products created overseas are restricted by the government, *shanzhai* then becomes a type of struggle against the anti-globalisation policy that prevents people

17 Ibid.

18 Qiu, '*Shanzhai* Culture'; Sheng Zhu and Yongjiang Shi, '*Shanzhai* Manufacturing: An Alternative Innovation Phenomenon in China: Its Value Chain and Implications for Chinese Science and Technology Policies', *Journal of Science and Technology Policy in China* 1.1 (2010): 29-49; Wu, '*Shanzhai* Economy'.

19 William Hennessey, 'Deconstructing *Shanzhai*-China's Copycat Counterculture: Catch Me If You Can', *Campbell L. Rev.* 34 (2011): 617.

20 Zhang and Fung, 'The Myth of "Shanzhai" Culture'.

21 Ho, '*ShanZhai*'.

them from consuming the foreign products they like. Essentially, *shanzhai*, or copycatting practice, occurs when people cannot consume certain products because they are restricted by either price, or policy. Responding to the copying and creation of such products, *shanzhai* becomes a reproductive innovation that produces 'foreign' products locally. Nevertheless, as argued earlier, although a related concept, 'second-degree reproduction' should not simply be equated to *shanzhai* as it involves a more complex relationship between the reproduced and the original product. If we understand *shanzhai* as a creative grass-roots parody of a particular brand or product, then 'second-degree reproduction' claims more productive power to internalise and localise the product within its own creativity. In other words, a *shanzhai* product exists beyond a comparison of the original product and itself. One can enjoy the *shanzhai* Spring Festival Gala because he or she knows the original Gala show produced by CCTV. However, a second-degree reproduced product is usually an independent product, unknown to its audience and with a less obvious connection to the original product. Because of the government's cultural policies, foreign products are mostly blocked in the country. As consequence, one might not even know *Niconico* or AKB48 before they know Bilibili or SNH48 but can still enjoy the products without any knowledge of their origins. In the case of second-degree reproduction, the imitated product does not speak to its origin, as in *shanzhai* culture, but speaks directly to, first, its audience, who cannot engage with certain foreign cultures normally; and second, the 'mainstream' discourse that blocks these cultures.

Subcultural Space and Second-degree Reproduction

Echoing *shanzha*i, the second-degree reproduction that occurs in the subcultural and Japanese anime fan community reflects how Japanese popular culture is generated in Chinese youth's own cultural practices and production. From genres to language use, the second-degree reproduction conducted by anime fans basically follows every tradition in Japanese ACG (anime, comic, and games) culture. Most of the jargon and slang words related to second-degree reproduction have their origins in Japanese fandom. Thanks to information technology and social networks, many of these vocabularies have transcended the limits of anime fans and are now popular among all youths who are familiar with internet culture. A typical example is the widespread use of 宅 ('geek') and 萌 ('cute') online. The former term is developed from the Japanese word 御宅 (otaku), which literally means 'people who only stay at home'. In ACG culture, otaku refers to those who are addicted to the virtual animated world. The latter word 萌 also originates from Japanese and the word 萌え (moe), which is used as an adjective to describe the cult status of things. Moreover, Chinese female slash fans call themselves 腐女, which literally means 'rotten women'. This word is a straightforward loan from the Japanese word 腐女子 (fujoshi), which refers to the female fans who fantasise about beautiful male couples. When naming their recreation and re-production, Chinese fans would usually use the word 同人 (doujin, as pronounced in Japanese), which is also imported from the Japanese term 'どうじん', referring to 'the recreated work with the same characters'. Terms like *Tongren* use the same Han characters as in Japanese but are written using Chinese Pinyin instead of Japanese Pinyin. Under the umbrella of '同人', subcategories and subgenres of second-degree reproduction are also shaped by Japanese ACG culture (Table 1). By practicing within these genres, China's anime fans are producing Japanese-style cultural products following Japanese innovation patterns.

These terms, which are based on Japanese ACG culture, suggest that the format of second-degree reproduction in Chinese anime fandom is a duplication of Japanese fandom, where fans draw Japan-style pictures, sing Japanese songs, and dance in sailor skirts or Japanese kimonos, but, at the same time, they target these reproductions as 国人作品 ('Chinese fan-work'). Instead of introducing 'exotic' Japanese fan culture and fan work, Chinese fans create their own reproductions with clear Japanese features and that are cultivated in their daily cultural practices and consumptions. In the same way that '*shanzhai*' enables poor people to use self-made, iPhone-like mobiles when iPhones are too expensive for them, this kind of reproduction allows Chinese anime fans to produce and consume Chinese-made but Japan-style contents when the original Japanese anime are strictly controlled by the government.

Many of the important online spaces that eventually become the 'mountain village' for youths to conduct and consume second-degree reproductions are the product of such reproduction as well. One example is the emergence of instant commentary directly shown on the video screen (Danmu, in Chinese), one of the most popular and commercially successful digital subcultural spaces embraced by youth recently. As mentioned earlier, the availability of technology and platforms has enhanced the level of participation from purely discussions to second-degree recreation. These video websites, with a built-in simultaneous commenting system, allow viewers to upload self-made videos and share their comments at any moment during the video play. Once a comment has been posted, it is shown on the screen for anyone to see, as long

2.1.1 Slang words translated from Japanese ACG culture

Japanese term	Chinese term	Meaning in Chinese context
どうじん（Douiin）	同人 (Tong-ren)	Fan's reproduction with elements (character, worldview etc.) in original animation. Genres include fan-fic, fan-pic, cover song etc.
どうじんし (Doujinshi)	同人志 (Tong-ren-zhi)	Fan magazine carrying Tong ren content.
舞ってみた (Wattemita)	舞见 (Wu-jian)	Fan-dancers who post video of their imitated dancing of anime-related dance.
歌ってみた (Utattemita)	唱见 (Chang-jian)	Fan-singers who post their cover songs of anime-related songs.
突っ込み (Tsukkomi)	吐槽 (Tu-cao)	To tease and to make fun of something or someone.
マッド (Maddo)	MAD	Music Anime Donga, referring to the re-editing anime clips to produce a relatively complete fan-made video with its own theme or narrative.
鬼畜 (Kichiku)	鬼畜 (Gui-chu)	A specific genre in which a song is made by completely splitting and re-editing voice sources collected from various audio or video.
周边 (Shuuhenn)	周边 (Zhou-bian)	Comic and anime related peripherals.
声优 (Seiyu)	声优 (Sheng-you)	Vocal actors.

as they are at the same position. The two largest Danmu websites in China are *ACFun* and *Bilibili*, both of which have adopted the design and technology from the Japanese ACG Danmu website ニコニコ動画 (*Niconico Donga*). On the websites, animation fans post and repost episodes of Japanese animations, Korean reality shows, and American TV dramas that have already been translated by fan-sub communities. Users can also post self-edited videos such as MADs and MVs (music anime dōga's and music videos), which are created based on the episodes they love. In each video, fans discuss the video content with each other using the simultaneous commenting system. Each of these videos can thus be regarded as a small, subcultural space where fans can enjoy, communicate, and even self-make their favourite videos.

Similar to *Niconico*, the most common categories of self-post videos on *ACFun* and *Bilibili* include MAD, fan singing, fan dancing, and a gaming stream. The original texts they refer to might be diverse, but in most cases are globalised content. The reproductions are, like many other fan activities, motivated by affection for the genre. This makes second-degree reproductions very self-reflexive and entertaining and, more importantly, non-political.[22] With awareness of censorship and political restrictions even online, Chinese youths avoid mentioning any political subjects in their daily cultural practices. On websites such as *Bilibili*, any political discussions, whether or not they are noticed by a government department, are deleted immediately by the website administrators. In other fan-based communities, such as discussion boards or in QQ groups, there is always a clear rule forbidding political discussions. The best way for these fans to protect their own space and autonomy is to avoid political issues. These youth cultural practices are thus purely entertaining. The self-censorship, in this case, should be seen as a compromise to maintain as much autonomy as possible. This means that these marginal spaces have become habitats that cultivate a new sense of ideology that might be different from the official discourse, but that is far from radical and mature.

Although there is a complex relationship between China and Japan in terms of history and politics, the government seems to care less about the possibility that Chinese youth may be politically fond of Japan and its culture. The Otaku culture is purely non-political from the perspective of both officials and many of the fans. They would love to watch *APH*, a Japanese-produced animation depicting World War Two that turns the nations into comic characters, but does not link 'Japan' to the real Japan in history.[23] Steering away from politics has successfully protected the local space from official control and has enabled the subcultural community to enjoy a foreign cultural product relatively safely. At the same time, websites such as *Bilibili* have established member examination to ensure the 'purity' of their members. Members must answer a quiz with over 100 questions about ACG culture in order to gain permission to use the commentary system and watch some of the videos. Such a mechanism further ensures the de-politicisation of the websites, because only members with 'plenty of love' are allowed to conduct second-degree reproductions. It makes the community a purer and tighter one, in which everyone wants to protect it from possible restrictions from outside.

To a certain extent, structures like this set a boundary between the subcultural community and the real world. What youths are practicing and

22 Yiyi Yin and Anthony Fung, 'Youth Online Cultural Participation and Bilibili: An Alternative Form of Digital Democracy in China?', in *Media Integration for Participatory Democracy*, eds. Rocci Luppicini and Rachel Baarda (Hershey, PA: IGI Global, 2017).

23 Jenkins, 'The Cultural Context of Chinese Fan Culture: An Interview with Xiqing Zheng', Confessions of an ACA-fan (webpage), January 31, 2013, accessed June 18, 2016, http://henryjenkins.org/2013/02/the-cultural-context-of-chinese-fan-culture-an-interview-with-xiqing-zheng-part-one.html.

consuming are thus from this subcultural world instead of from the official media culture of China. In other words, this world informs youths about things they won't get from the mainstream discourse, and it encourages them to disseminate it. *Griever of Ideology* is a case in point.[24] In 2016, a girl posted on *Weibo* about her experience of being sexually assaulted in Beijing. A huge discussion about this issue and feminism appeared on *Bilibili* afterwards following the posting of a fan-made ballad called *Griever of Ideology*. Till now, feminism has not been a common subject in Chinese official discourse. Several famous feminist *Weibo* accounts have even been shut down by the authorities in recent years. Nevertheless, the discussion on *Bilibili* survives because of the subcultural 'cover' of the *Griever of Ideology*. This song is composed in the trip-hop genre and features Japanese-style animation. The lyrics are ironic, but without any direct complaints or reference to the social issue *per se*. At first glance, the song is an ordinary work that is common on subcultural websites like *Bilibili*, but, in fact, it has become a forum for youths to learn, elaborate, and discuss feminism, a subject that is not seriously discussed in Chinese reality. This case shows how subcultural practices embrace different ideologies within their relatively marginal space. In the following case studies, the chapter argues how this virtual and subcultural space, with reproduced products, cultivates certain ideologies that are not encouraged or mentioned in China's mainstream discourse.

This world, on the one hand, is a secret base surrounded by foreign cultural products for fans to enjoy. It becomes a kind of utopia that is interesting, entertaining, and cannot be found in the 'boring' reality where the desirable contents are banned. On the other hand, the world is shaped and maintained not only by foreign products, but also by the youth's own cultural practices, namely second-degree reproductions. By continually recomposing existing texts, the reproductions maintain the local spaces as a productive and autonomous system in which youths replicate sources, not only from cultural products, but also from daily life. In practice, they are building their imaginations into a virtual or second-dimensional world that is deeply influenced by foreign culture, especially Japanese Otaku culture.

From Being Globalised to a Globalisation Strategy

What second-degree reproduction implies is not just a copycatting form of grass-roots creativity among Chinese youth, it is a hint to rethink how

24 Yin and Fung, 'Youth Online Culture Participation'.

China, as a country with strict control of imported foreign products, is transitioning towards an alternative form of globalisation. Instead of being globalised by the foreign culture coming from the 'centre', as the structural imperialism theory would suggest, globalisation proceeds from the inside out, as a self-generated and domestic process.[25] As a developing country with a relatively weak tradition of cultural innovation and so-called soft power, China is commonly considered as the cultural periphery in both Asia and on a global scale, where Western products and Japanese or Korean pop waves invade the domestic market.[26] This post-colonial perspective provides a valuable pattern for examining China's status in globalisation after the policy of Reform and Opening in the late 1970s. During this period, foreign cultural products, such as Hollywood films and Korean TV dramas, were imported into the country. Whereas in recent decades, as the anti-globalisation policy was established by the Chinese government as a kind of resistance to 'Westernisation', a domestic youth culture embracing the practices of second-degree reproduction, emerged as an internally globalisation-sought strategy to play with global culture in local production. This process, different from traditional understanding of localisation which refers to contextualising of the foreign product, is domestically developed by cloning global culture into local production.

The interpenetration and two-way dynamic here reflects the complex contradiction between official anti-globalisation ideology and the non-official desire to consume global contents. At this point, the second-degree reproduction actually creates a relatively safer space where global cultures engaging in cultural consumption are not really 'global'. There are, for sure, potential contradictions between the ideology embedded in foreign popular culture and China's own ideology. Nevertheless, by avoiding politically sensitive issues, the domestic youth culture seems to be no threat to the state and so there is no reason to ban it. The subcultural space has thus become a grey area that is tacitly approved by the government. The government even tends to support this kind of local globalisation to develop the local creative industry and to increase the soft power of the nation.

One interesting case of such internal global content is the initiation of SNH48, a female idol group popular among Chinese youths especially in cities like Shanghai. As one can speculate from the name, SNH 48 is

25 Johan Galtung, 'A Structural Theory of Imperialism', *Journal of peace research* (1971): 81-117.
26 Arif Dirlik, 'The Postcolonial Aura: Third World Criticism in the Age of Global Capitalism', *Critical inquiry* 20.2 (1994): 328-56; Yuezhi Zhao and Paula Chakravartty, *Global Communications: Toward a Transcultural Political Economy* (Lanham, MD: Rowman & Littlefield Publishers, 2007).

basically a replica of the Japanese idol group AKB48. Initially, the two groups were co-sponsored by the same Japanese corporation. In 2016, SNH48 was independent from the Japanese corporation and was produced and developed by a local entertainment company in Shanghai. As the Chinese version of AKB48, almost every aspect of SNH48 is similar to the Japanese group, from the group's design, the music genre, and their stage performances. They have produced a lot of replications and secondary productions of AKB48's previous works, including their songs, dancing, and music videos. The costumes are almost the same too and, in many cases, they are very Japanese in terms of style and design. For one of their albums, SNH48 produced a music video where the girls dressed in Japanese high-school uniforms. Since most Chinese students wear sports uniforms in school, the sailor-style uniforms were a clear representation of Japanese rather than Chinese school culture. In another case, the Chinese girls dressed in bikinis, which is also copied from AKB48, dancing on the beach.

Such performances are rare in Chinese domestic productions, and while not totally unacceptable, this Westernised and sexy design is very different to traditional Chinese culture and values. As a domestic Shanghai idol group, what has been reflected in SNH48's works is not Shanghainese or even Chinese, but Japanese youth culture. From music to performance, to the idol selection strategy, SNH48 copies everything from AKB48. In the same way as anime fans consume Japan ACG culture on *Bilibili*, SNH48 should be seen as another example of 'global' content created domestically.

In addition to copycatting Japanese culture in local reproductions, Korean popular culture is also adapted and used in the Chinese domestic idol-training system. For instance, TFBoys, the most popular youth idol group in mainland China in the past three years, is a successful replication of K-pop and the Korean star-production system. The idol group was formed by three male teenagers who have been trained in entertainment from a young age. After several years of training, the boys were presented to the public via their singing and dancing as well as a self-made reality show that was released online. Although the boys are extremely local – they even have Chongqing accents – the idol group copies everything from the Korean idol system: The boys were selected from the 'trainee' system, of the type that is typical to the Korean entertainment industry. Many of their songs and music videos were produced in Korea or by a Korean production team. The group's design is also similar to Korean male groups. The fact that TFBoys's crazy popularity actually started with the groups first hit on *Bilibili* testifies to how this kind of Korean format has been embraced by China's youth. Again, Korean culture has not been introduced to the Chinese public by importing Korean idol groups or stars; rather, it is embedded in the domestically produced idol groups through the practice of *shanzhai* or second-degree reproduction. In terms of global culture, TFBoys should be considered as another case that internally clones foreign culture in its own market.

However, to keep this internal globalisation safe from the government's control, localisation is necessary in the 'shanzhai-ing' process. Anything that is sensitive to the dominant power has to be removed. For instance, the TFBoys are extremely careful to avoid any behaviours that might be considered as unhealthy for teenagers. The boys are trained to be very polite and hard-working, acting like the best students in the state. In order to not lead other teenagers astray, they must go to school and behave well. For this reason, they rarely sing love songs. Instead, by being sunny and positive, the TFBoys are accepted by the mainstream as good role models for Chinese youth. Like SNH48, the success of the TFBoys can also be seen as a celebration of youth culture in which the globalised culture, or at least the regional culture in Asia, has been developed locally within China.

Evidence also suggests that even some more radical and somehow dangerous replications of foreign culture can also survive as long as they avoid politics. The voting system used to rate SNH48, which was initially copied from AKB48, is an example. Like the voting system developed by the Japanese, every year SNH48 holds an anniversary vote encouraging fans to buy their albums and vote for their favourite members. Based on the results, the top seven members with most votes become the focus of all commercial activities in the following year. Before the final vote, a series of campaigns are held during the weekends. In Japan, the top singers in AKB48 receive around 7-20,000 votes per year. Of course, the political voting system and relevant democratic mechanism are very sensitive to Chinese officials. To avoid ideological censorship, SNH48 came up with a series of strategies to make the voting as non-political as possible. For example, the title was changed from the more political 总选举 ('final election') to the term 总决选 ('final decision'). The format of the campaign was also modified. Consequently, it avoids attention from the authorities by emphasising its non-political nature: It is just a commercial engagement, not a political campaign at all.

As new business models such as those employed by the TFBoys and SNH48 emerge, what could such developments mean for youth culture? As discussed above, its non-political nature protects it as a safe and marginalised subculture that is far from the government's purview. However, as capital is involved, the subculture has started to step into the public sphere. As the most popular idol groups in China, neither SNH48, nor TFBoys can be seen as completely trivial or marginalised. Along with the process of commercialisation, the subcultural space is now gradually morphing into a mass cultural space. At this stage, the youth culture must negotiate with the dominant power. In the past two years, websites like *Bilibili* and *ACFun* have continuously changed their own policies for governing video and comment content. Groups like TFBoys and SNH48 have also tried to present some 'mainstream' songs. By being politically correct, youth culture is attempting to protect its autonomy.

New Derivatives: Imported Global Culture Manufactured

As second-degree reproductions like the TFBoys and *Bilibili* become popular in China, capital, both local and global, is playing an increasingly significant role in this process. Although trivial and marginal in the eyes of government, the virtual world has developed into an attractive investment. As more and more youths devote themselves to the subcultural space and cultural practices such as second-degree reproductions, they also become important target consumers in the domestic market. Advertisements target youth under 25 and pay attention to the ACG culture. Indeed, ever more advertisers consider *ACFun* and *Bilibili* as ideal platforms, because they are websites where young people gather. Till 2015, *Bilibili* had attracted more than 50 million users. More than 75 per cent of them are youth under the age of 24. Benefitting from ads and other commercial collaborations, the company Tencent estimates *Bilibili* has reached a value of 1.5 billion RMB in 2015.[27] The fact that youth are generating a lot of revenues based on Japanese animations and other globalised contents has pushed the industry to look into the contents that they find desirable.

To further explore the capital value of this internal global culture, a new trend has emerged as a calculated strategy for manufacturing global culture outside China that can then be imported back into the country. Instead of passively adapting popular global elements into local investments, this new strategy of the Chinese creative industries allows a certain degree of contra-cultural flow in the form of investing and participating in global production. For instance, two animations, 从前有座灵剑山 (*'Reikenzan'*) and 雏蜂 ('Bee'), were imported to China in 2015 and 2016. The two animations were both adapted from manga produced by Chinese authors, which were

27 Techweb, 'The Bullet-comment Website Bilibili is Invested by Tencent with 1.5 Billion Market Valuation', *NetEase*, November 9, 2015, accessed September 4, 2016, http://tech.163.com/15/1109/14/B802872P000915BF.html.

completely Japanese-style in terms of their genre, characters, narratives, and drawing style. As typical cases of second-degree reproductions, these domestic manga were proposed to Japanese animation corporations as a collaborative project with China. As a consequence, the foreign-style animations were allowed to be imported into the Chinese animation market because they were originally made by Chinese authors.

As most globalisation studies argue, inequality among nations and regions leads to structural cultural flow from the centre to the periphery. The aforementioned strategy actually proposes the possibility to engage in a circulation where certain culture moves from the periphery to the centre and then flows back to the periphery again.[28] By investing capital in the entire trajectory of circulation and distribution, there is, as shown in many Hollywood blockbusters, such as *Ironman 3* and *Transformers 4*, a chance for a reverse cultural flow occurring in global culture manufactured at least partly by Chinese industry.[29] As cases like *Bee* imply, the next step after the *shanzhai*-ing of global culture, is to merge this domestically produced 'foreign' culture into the global innovation culture, making it appropriate to be imported or reproduced again in China.

Potential Contradictions: Ideal World vs. Reality

As stated above, this new form of globalisation is relatively 'safe' because it makes itself marginal by replicating foreign popular culture locally, and more importantly, by erasing any political traces that might be embodied in foreign cultural products. However, being non-political and of no interest to the government does not mean that this kind of alternative globalisation has no cultural and social impact in China. On the contrary, it is the non-political that has sheltered the potential social power embedded in specific cultural practices so that the current youth culture might have a long-term impact on Chinese society.

Firstly, this kind of second-degree reproduction per se has become a challenge against the anti-globalisation policies set by the country. Blooming in local spaces, foreign cultural products have merged into cultural life,

28 Galtung, 'A Structural Theory of Imperialism'; Huaiting Wu and Joseph Man Chan, 'Globalizing Chinese Martial Arts Cinema: The Global-Local alliance and the Production of Crouching Tiger, Hidden Dragon', *Media, Culture & Society* 29.2 (2007): 195-217.

29 Richard A. Rogers, 'From Cultural Exchange to Transculturation: A Review and Reconceptualization of Cultural Appropriation', *Communication Theory* 16.4 (2006): 474-503.

especially among the youth. Whether it is a signal of cultural homogeny or cultural hybridisation is beyond the scope of this chapter, but what is certain is that the cultural flow has penetrated into the domestic market. Having access to mass media products that are banned by the state means that youths are still able to choose what they'd like to receive, despite the pressure of the dominant ideology.

Furthermore, the contradiction between youth culture and the anti-globalisation policy parallels with another potential contradiction that exists between the youth's imagined world versus the reality. Simply put, similar to what cultivation theory has suggested, a long-term exposure to foreign products may influence subsequent generations' own ideology, even though the contents themselves are non-political. These ideological changes then shape their imagination of an ideal world that is, in some cases, significantly at odds with the dominant ideology in China. For example, the anime *Attack of Titan*, despite its implication of militarism, portrays a world where characters imprisoned by the authorities (in this case, 'the wall') struggle and sacrifice in a fight against the undefeatable power (the Titans). Depicting how characters fight together everyday against the Titans, the theme of the animation is overthrowing the controlling authority and power and finally reaching a free world. Another animation, *Naruto*, emphasises the appeal of fighting for a peaceful and equal world where power is used to protect rather than control people. Both animations, in fact, are creating an unreal and idealised world where the dominant power or authority is portrayed as evil or a barrier. In the Chinese reality, the authority of the party and the state clearly does not welcome questioning or resistance. Influenced by the ideal created by these global products, youths are more likely to be sensitised to aspects of individual equality, human rights, and democracy, ideas that the Chinese government find challenging.

Similarly, fans of SNH48 are familiar with the relatively democratic civic power instilled by the group's campaign and voting system, mentioned earlier. In the process of ranking and electing a 'top girl' for SNH48, fans absorb the democratic habits of selecting and supporting their own leaders. As the voting is held publicly on WeChat, a highly popular social media platform in China, this kind of participatory democracy then moves beyond the subcultural local space to a very public space. It also implies the potential of youth culture to become a kind of social power that may challenge the dominant ideology. When subculture becomes a part of daily life, the youth learn rules and ideology from the desirable products they are exposed to. Admittedly, as Meng has argued when discussing the reality show *Super Girl*, this kind of 'democratic' practice is far from a radical challenge to the existing social order and authority.[30] However, it does not deny the progressive value of practicing certain behaviours such as voting and campaigning, especially when *Super Girl*, a reality show aired on TV, has been largely limited and censored by the authority, while cases like SNH48 remain largely subcultural products. But clearly, the SNH48 subculture is growing as replicas in the form of BEJ48 and GNZ48 in North China and South China, respectively, have now emerged.

The youth not only find certain social rules and ideology in their ideal world, but also their own social models. The new generation no longer regard models seen in the mass media, such as Chris Li – a pop idol – or political prototypes promoted by the state, such as Leifeng – a communist legend – as society's political leaders. Now, growing up with the internet, youth actively search for ideal types or models that are specifically relevant to them; consequently, there are fewer 'big idols' than there were for the previous generation.[31] In the same way as they regard the animated world as the ideal, many youths also find the virtual characters in animations relevant, charismatic, and desirable. They learn not from models in reality, but in the animation, which might affect their attitude towards so-called real practices. A noticeable theme in many popular Japanese manga and anime, especially in certain genres such as *Shonen* ('youth') manga, is adventures in which the character fights against strong and powerful enemies. This influences the youth in terms of how they understand community, social relationships, class, and power. While the government, or the 'adults' consider animations and other subcultures as trivial and childish play, the youth are actually building their own world based on these productions.

Conclusion: Producing Youth Culture

All these potential consequences of long-term practices imply the possibility that the current youth culture in China may become a significant social power in the future. There is a non-political but still rebellious and subcultural challenge posed towards the dominant ideology. By making sense of the ideal world, the youth have their own rules, policies, and worldviews

30 Bingchun Meng, 'Who Needs Democracy if We Can Pick Our Favorite Girl? Super Girl as Media Spectacle', *Chinese Journal of Communication* 2.3 (2009): 257-72.

31 The State Council Information Office of the People's Republic of China, 'How Did Idolization of Chinese Youth Change', 2015, accessed in June 20, 2017, http://www.scio.gov.cn/zhzc/2/32764/document/1433818/1433818.htm.

that originate not in the 'real' Chinese society, but in the virtual and ideal world they have imagined. As the state bans foreign content that has political tendencies, the youth are actually practicing their desirable and globalised subculture locally within China through *shanzhai* practices.

This chapter has illustrated a new form of globalisation that is self-generated and domestically produced. It also examines how *shanzhai* culture or second-degree reproduction encourages such alternative forms of globalisation in the complex dynamic between commercial culture and the nation's anti-globalisation policy. As the chapter argues, second-degree reproduction has created a new youth culture in which discourses that potentially contradict the state ideology are emerging and developing in their own local spaces. Second-degree reproduction currently remains subcultural, if not underground, but as the rhizomes penetrate via the internet, subcultures grow and are starting to surface in all major cities. As these subcultures spread across the Chinese territories, it may not be long before the state or mainstream media recognise that these subcultures are the mutations of today's youth culture.

This research project was funded by a grant from the Research Grant Council of HKSAR (Project no. GRF 14617716) and the Key Fund of the National Social Science Foundation of China: Arts category (18ZD12).

Bibliography

Administration of Culture of the PRC, '文化部首次公布网络动漫产品"黑名单"', 2015. Accessed March 10, 2019. http://www.cac.gov.cn/2015-06/08/m_1115549876.htm.

Arnett, Jeffrey Jensen. 'The Psychology of Globalization', *American psychologist* 57.10 (2002): 774-83.

CICAF. 'Introduction of the Chinese International Comic & Animation Festival (CICAF 2015).' Accessed August 28, 2016. http://www.cicaf.com/xwzx/content/2015-01/01/content_5576429.htm.

Dirlik, Arif. 'The Postcolonial Aura: Third World Criticism in the Age of Global Capitalism', *Critical inquiry* 20.2 (1994): 328-56.

Galtung, Johan. 'A Structural Theory of Imperialism', *Journal of peace research* (1971): 81-117.

Gong, Haomin, and Xin Yang. 'Digitized Parody: The Politics of *Egao* in Contemporary China', *China Information* 24.1 (2010): 3-26.

Guan, Zhong-Afu. 'Where is the Path? How Can Japanese Animation Step on the Chinese Road?', *Insights*, April 29, 2015. Accessed September 4, 2016. http://culture.ifeng.com/insight/special/anime/.

Hennessey, William. 'Deconstructing *Shanzhai*-China's Copycat Counterculture: Catch Me If You Can', *Campbell L. Rev.* 34 (2011): 609-61.

Ho, Josephine. '*ShanZhai*: Economic/Cultural Production through the Cracks of Globalization'. Plenary presentation (Hong Kong: ACS Crossroads: 2010 Cultural Studies Conference, 2010).

Jenkins, Henry. 'The Cultural Context of Chinese Fan Culture: An Interview with Xiqing Zheng', Confessions of an ACA-fan (webpage), January 31, 2013. Accessed June 18, 2016. http://henryjenkins.org/2013/02/the-cultural-context-of-chinese-fan-culture-an-interview-with-xiqing-zheng-part-one.html.

—. *Textual Poachers: Television Fans & Participatory Culture* (New York: Routledge, 1992).

Meng, Bingchun. 'Who Needs Democracy if We Can Pick Our Favorite Girl? Super Girl as Media Spectacle', *Chinese journal of communication* 2.3 (2009): 257-72.

Ministry of the Culture of the People's Republic of China, 'The Conference of Censoring the 23rd Group of Illegal Internet Cultural Activities', 2015. Accessed September 4, 2016. http://www.mcprc.gov.cn/whzx/whyw/201503/t20150331_439865.html.

Qiu, Lin-chuan. '*Shanzhai* Culture in the Era of the Internet', *21 Century* 112 (2009): 121-39.

Rogers, Richard A. 'From Cultural Exchange to Transculturation: A Review and Reconceptualization of Cultural Appropriation', *Communication Theory* 16.4 (2006): 474-503.

Sonna, '中国 ACG 圈的发展历史是怎样的，有哪些重要网站', Zhihu 2004. Accessed March 10, 2019. https://www.zhihu.com/question/22561605/answer/25534694

The State Council Information Office of the People's Republic of China. 'How Did Idolization of Chinese Youth Change', 2015. Accessed in June 20, 2017. http://www.scio.gov.cn/zhzc/2/32764/document/1433818/1433818.htm.

Techweb. 'The Bullet-comment Website Bilibili is Invested by Tencent with 1.5 Billion of Market Valuation', *NetEase*, November 9, 2015. Accessed September 4, 2016. http://tech.163.com/15/1109/14/B802872P000915BF.html.

Wu, Huaiting, and Joseph Man Chan. 'Globalizing Chinese Martial Arts Cinema: The Global-Local Alliance and the Production of Crouching Tiger, Hidden Dragon', *Media, Culture & Society* 29.2 (2007): 195-217.

Wu, Jing. '*Shanzhai* Economy and New Media Creativity: A discussion of the Formation of Postmodernist Public Sphere', *China Media Report*, 2010. Accessed August 28, 2016. http://www.chnmedia.com/wordpress/?p=2578.

Xinhuanet. 'New Rules by Administration of Broadcasting and Television: Foreign Animations are Forbidden to Air on TV during Golden Period', *Xinhuanet*, 2006.

Accessed September 4, 2016. http://news.xinhuanet.com/politics/2006-08/13/content_4954807.htm.
Yin, Yiyi, and Anthony Fung. 'Youth Online Cultural Participation and Bilibili: An Alternative Form of Digital Democracy in China?'. In *Media Integration for Participatory Democracy*, eds. Rocci Luppicini and Rachel Baarda (Hershey, PA: IGI Global, 2017).
Yu-sheng. 'The Ministry of Culture Announces the Animation Blacklist: Attack on Titan is Included', *China News*, June 8, 2015. Accessed September 4, 2016. http://www.chinanews.com/cul/2015/06-08/7329965.shtml.
Zhang, Lin, and Anthony Fung. 'The Myth of "Shanzhai" Culture and the Paradox of Digital Democracy in China', *Inter-Asia Cultural Studies* 14.3 (2013): 401-16.
Zhao, Yuezhi, and Paula Chakravartty. *Global Communications: Toward a Transcultural Political Economy* (Lanham, MD: Rowman & Littlefield Publishers, 2007).
Zhu, Sheng, and Yongjiang Shi. '*Shanzhai* Manufacturing: An Alternative Innovation Phenomenon in China: Its Value Chain and Implications for Chinese Science and Technology Policies', *Journal of Science and Technology Policy in China* 1.1 (2010): 29-49.

About the Authors

Anthony Y.H. Fung is Professor in the School of Journalism and Communication at the Chinese University of Hong Kong. He is also Professor in the School of Art and Communication at Beijing Normal University under the Global Talents Scheme. His research interests and teaching focus on popular culture and cultural studies, gender and youth identity, cultural industries and policy, and new media studies. He authored and edited more than ten Chinese and English books. His recent books are *Youth Cultures in China* (Polity Press, 2016) (co-authored with de Kloet) and *Global Game Industries and Cultural Policy* (Palgrave Macmillan, 2017).

Yiyi Yin is a Ph.D. candidate at The Chinese University of Hong Kong. Her research interests include fandom studies, popular culture and cinema studies. From a cultural-historical perspective, her current work focuses particularly on the evolution of fan practice and fan subjectivity in China.

2.2 Creative '*Shanzhai* Labour'?

Leung Mee-ping's 'Made in Hong Kong/Shenzhen'

Louis Ho

Abstract

Shanzhai is a term that encompasses a wide range of concepts including copy-and-paste, grass-roots cleverness, and even anti-authoritarian culture. In traditional Chinese, *shanzhai* refers to 'villages in the mountain with stockade houses' and, in Shenzhen, there is a real village called Dafen Oil Painting Village, which is famous for making '*shanzhai* art' – replicas of oil paintings by globally renowned artists. The Hong Kong artist Leung Mee-ping carried out a project called 'Made in Hong Kong/ Shenzhen', in which the artist produced a series of 'Hong Kong-themed' paintings with the 'techniques' she learnt at the Dafen Village. This chapter attempts to explore the cultural dynamics of Leung's project by rethinking various related notions, including authorship, creativity, *shanzhai*, and appropriation art.

Keywords: *shanzhai*, shanzhai art, creative labour, craft labour, Dafen Village

Introduction

In 1998, one year after the handover, the Hong Kong government announced a proposal to construct the West Kowloon Cultural District.[1] After a series of slogans, debates, controversies, and propaganda,[2] a development plan

1 *South China Morning Post*, '1998 Former Chief Executive Tung Chee-hwa Announces Plan to Develop the West Kowloon Cultural District', September 30, 2011, p. 3. See also West Kowloon Cultural District Authority (webpage), assessed October 1, 2016, http://www.westkowloon.hk/.
2 Chi-fai Cheung, 'Planners Throw Out Arts Hub Objections Board Confirms Zoning Blueprint Giving West Kowloon Developer a Free Hand', *South China Morning Post*, December 13, 2003, p.1;

Kloet, Jeroen de, Chow Yiu Fai, and Lena Scheen (eds), *Boredom,* Shanzhai, *and Digitisation in the Time of Creative China*. Taylor & Francis Group, 2019
DOI: 10.5117/9789462984745_CH2.2

for the West Kowloon Cultural District was finally approved by the Chief Executive and the Executive Council in 2013.[3] From initial planning to the later construction process, the *M-plus* (M+), a new museum for visual culture has been a the centre this much anticipated and discussed project. The museum will be built with an estimated budget of 4.9 billion.[4] It will feature a cross-media gallery for exhibiting art, design, and architecture and will include images from the twentieth and 21st century.[5] The key exhibits in the gallery have been provided by Dr. Uli Sigg, a renowned Swiss art collector whose collection comprises 1,510 artworks created by over 300 contemporary artists.[6] In 2016, while the Main Gallery of M+ was still being built, the M+ curatorial group organised an exhibition entitled 'M+ Sigg Collection Exhibition: Four Decades of Chinese Contemporary Art' at ArtisTree, a privately owned art space showcasing 80 pieces of art by 50 contemporary artists in China.[7] By showing this world-renowned work, the event attempted to illustrate the development history of contemporary art in China. While visiting the exhibition, I wondered how the delicate relationship between Hong Kong and China was manifested through this first gallery exhibition of M+, one of the most iconic cultural venues since the handover. Interestingly, while looking at paintings by Fang Lijun, Zhang Xiaogang, and Zeng Fanzhi, an uncanny feeling suddenly embraced me. These masterpieces, while symbolising the rise of contemporary Chinese art in the international arena, have actually flooded the market for replicated paintings, filled the internet, and they have been used as images on different consumer goods. Facing the original works that represented the creativity

Tai-luk Lui, 'City-branding Without Content: Hong Kong's Aborted West Kowloon Mega-project, 1998-2006', *International Development Planning Review* 0.3 (2008): 215-26; Joyce Ng, 'Arts Hub Seeks More Cash for Basement CEO of West Kowloon Cultural District Authority Says It Needs Extra Government Money in Coming Months to Build a Vast Underground Complex', *South China Morning Post*, July 26, 2012, p. 2.

3 Legislative Council of HKSAR, *Panel on Home Affairs and Panel on Development Joint Subcommittee to Monitor the Implementation of the West Kowloon Cultural District Project: The approved Development plan of the west Kowloon cultural district*) [LC Paper No.CB(2)658/12-13(01)] (HK: HKSAR, 2013a), assessed October 1, 2016, http://www.legco.gov.hk/yr12-13/english/panels/wkcd/papers/wkcd0225cb2-658-1-e.pdf.

4 Legislative Council of HKSAR, Legislative Council Subcommittee on West Kowloon Cultural District Development. *M+ in West Kowloon Cultural District* (HK: HKSAR, 2008), assessed October 1, 2016, http://www.legco.gov.hk/yr04-05/english/hc/sub_com/hs02/papers/hs020524wkcd-546-e.pdf.

5 Legislative Council of HKSAR, *The Study of the West Kowloon Cultural District Development Project* (HK: HKSAR, 2013b), assessed October 1, 2016, http://www.legco.gov.hk/yr1213/english/panels/wkcd/papers/wkcd0225cb2-658-1-e.pdf.

6 See West Kowloon Cultural District Authority (webpage), http://www.westkowloon.hk/.

7 Ibid.

and artistic talents of Chinese artists, I wondered how to make sense of these works using the concepts of creativity, *shanzhai* ('fake stocks'), and artistic (re-)production.

By the 2000s, 'creativity' had become a popular discourse for policymakers and the public across many countries, including China; at the same time, critical discussions emerged about the *shanzhai* phenomenon.[8] Originally a derogatory term used to describe the practice of the Chinese copyist, the *shanzhai* phenomenon slowly entered popular culture and has subsequently expanded to mean a productive mechanism for the grass-roots culture to critique, resist, and fight against the culture of the mainstream elite class.[9] Considering 'creativity' and *shanzhai* as interwoven concepts, this study asks: how can we understand *shanzhai* as a form of production, and even creativity? Is there a hierarchy of creativity, in which specific forms of creativity are superior to the others? Which parameters can we use to distinguish 'superior' creativity? Through the textual analysis of a conceptual artwork by Hong Kong artist Leung Mee-ping entitled 'Made in Hong Kong/Shenzhen', which focuses on Dafen Oil Painting Village (the biggest replica painting production cluster in China), an in-depth, semi-structured interview with the artist, and empirical fieldwork in Dafen Oil Painting Village, this study explores and discusses the complexity inherent in the concept of creative labour within the domain of *Shanzhai* culture.

Located in the Longgang District of Shenzhen, Guangdong, Dafen Oil Painting Village has gradually become the 'Art Production Village' since the 1980s. In recent decades, the area has grown from a town of 200 people to a hub with over 8,000 painter-workers.[10] Initially, Dafen Oil Painting Village mainly focused on souvenir painting, which produced replicas of famous paintings by Western artists, such as Leonardo da Vinci, Rembrandt, and Van Gogh. Consequently, the *shanzhai* phenomenon has caught the attention of the Chinese government and media refer to Dafen Oil Painting Village as the production site of '*Shanzhai* Art'.[11] Owing to its unique production

8 Lin Zhang and Anthony Fung, 'The Myth of "*Shanzhai*" Culture and the Paradox of Digital Democracy in China', *Inter-Asia Cultural Studies* 14.3 (2013): 401-16; Fan Yang, *Faked in China: Nation Branding, Counterfeit Culture, and Globalization* (Bloomington, IN: Indiana University Press, 2015).

9 Yang, *Faked in China*; Winnie Wong, *Van Gogh on Demand: China and the Readymade* (Chicago, IL; London: University of Chicago Press, 2013).

10 *Al Jazeera*, 'Demand Grows for Chinese Fake Art – Al Jazeera English', 2010, accessed October 1, 2016, http://www.aljazeera.com/video/asia-pacific/2010/08/20108305501l812822.html.

11 *China Daily*, 'Shenzhen Shows Vitality, Dream with "Paintbrush" at Expo', August 26, 2010, assessed October 1, 2016, http://www.chinadaily.com.cn/bizchina/shenzhen30years/2010-08/26/content_11209299.htm.

mode and representation, Dafen Oil Painting Village has not only been the research subject of many scholars,[12] but it has also been the theme of several major artworks over the past years,[13] including Leung Mee-ping's art project 'Made in Hong Kong/Shenzhen'.

In order to explore the research questions mentioned above, this chapter discusses Leung Mee-ping's work 'Made in Hong Kong/Shenzhen', a conceptual art project in which the artist acted as a student-apprentice in a Dafen painter-master's workshop and created paintings of popular Hong Kong scenes. Apart from performing visual and textual analysis on the art products and relevant art critics, it will also focus on the above-mentioned core discussion topic, i.e. creative labour and, in particular, the distinction between artistic labour and craft labour. In addition, this chapter is based on in-depth interviews with Leung Mee-ping and field research in Dafen Oil Painting Village, where semi-structured interviews were conducted with painters in the village.

Apart from the high level of attention that Leung's work has gained from the arts[14] and academic[15] fields, another reason for choosing this work as the study subject for this research is because of the meticulous records that the artist kept during its creation process. The existing studies on Leung's work, which will be discussed in the later part of this chapter, mainly focus on the creative *output*, while this study focuses on the creative *process*, which involves various forms of creative labour. With the detailed information about the creation process provided by the artist, this chapter studies the

12 Youping Wen, *The Rise of Dafen Village* (China: Shenzhen: Haitian Press, 2007); Eva Kit Wah Man, 'Influence of Global Aesthetics on Chinese Aesthetics: The Adaptation of Moxie and the Case of Dafen Cun', *Contemporary Aesthetics* 11 (2013); Si-ming Li, Hung-ha Cheng, and Jun Wang, 'Making a Cultural Cluster in China: A Study of Dafen Oil Painting Village, Shenzhen', *Habitat International* 41 (1/2014): 156-64; Vivian Li, 'Original Imitations for Sale: Dafen and Artistic Commodification', *Art History* 37.4 (9/2014); Wong, *Van Gogh on Demand*.

13 See Wong, *Van Gogh on Demand*.

14 Sin-yue Leung, '「被」遺棄的記憶——專訪香港藝術家梁美萍', *All about Macau Media* [論盡媒體], April 24, 2014, assessed October 1 2016, http://www.inmediahk.net/node/1022480; Wai-sum Cheung, '冷靜與熱情之間：訪香港藝術家梁美萍，也談「展銷場：中國製造」', *Stand News*, June 23, 2015, assessed October 1, 2016, https://thestandnews.com/art/冷靜與熱情之間-訪香港藝術家梁美萍-也談-展銷場-中國製造/; *Headline Daily*, '藝術去留系列 藝術創作 展示香港另一面 重新發掘本土', November 27, 2015, assessed October 1, 2016, http://news.stheadline.com/dailynews/content_hk/2015/11/27/352899.asp.

15 Lai-kwan Pang, *Creativity and Its Discontents: China's Creative Industries and Intellectual Property Rights Offenses* (Durham, NC; London: Duke University Press, 2012); Winnie Wong, *Van Gogh on Demand: China and the Readymade* (Chicago, IL; London: University of Chicago Press, 2013); Ming-hoi Lai and Eva Kit Wah Man, 與香港藝術對話：*1980-2014* (HK: Joint Publishing, 2015).

work as a piece of text and dwells on the conceptual thinking that takes artistic creation, *shanzhai* culture, and politics into consideration.

Created or Made in Hong Kong/Shenzhen?

Fan Yang's work *Faked in China* (2015) provides a historical account and theoretical framework for *Shanzhai* culture in contemporary China. By focusing on post-socialist Chinese society in the first decade of the 21st century, Yang's study discusses the neo-liberal consumer identity and explores how Chinese people have struggled with the tension between the nation-branding campaign 'From Made in China to Created in China' and the emerging *shanzhai* phenomenon. In her book, Yang extracted a report from *China Daily* in 2008 to show the close relationship between *shanzhai* and contemporary Chinese society, which said:

> It's a cold Sunday morning. Mr. Phony turned off his hiPhone alarm, put on Kabba suits and Adidos shoes, grabbed a coffee from the KFG downstairs and came back in a hurry for the latest episode of the popular sit-down Ugly Wudi. No spelling mistakes above. Products imitating famous brands have not been uncommon around China for a couple of years – and now they have a unified name, a brand if you like, '*Shanzhai*'.[16]

Referring to Lash and Lury's theory of the 'thingification of culture',[17] Yang argues that *shanzhai* is the complex formation of the global cultural industry.[18] Its distinct cultural and historical attributes make *shanzhai* more than merely the replication of products. Yang's way of understanding *shanzhai*,[19] which is actually consistent with what many *shanzhai* studies suggest,[20] is important if we situate Dafen in *shanzhai* culture. Winnie Wong, for example, argues 'in Dafen Oil Painting Village [...] there has never been

16 Fan Yang, *Faked in China: Nation Branding, Counterfeit Culture, and Globalization* (Bloomington, IN: Indiana University Press, 2015), 91.

17 Scott Lash and Celia Lury, *Global Culture industry: The Mediation of Things* (Cambridge: Polity, 2007).

18 Yang, 'China's Fake Apple Store: Branded Space, Intellectual Property and the Global Culture Industry', *Theory, Culture and Society* 07 (2014); *Faked in China*.

19 Yang, *Faked in China*.

20 Zhang and Fung, 'The Myth of "*Shanzhai*" Culture'; Jeroen de Kloet and Chow Yiu Fai, '*Shanzhai* culture, Dafen art, and copyright'. In *Routledge Handbook of East Asian Popular Culture*, eds. Koichi Iwabuchi, Eva Stai and Chris Berry (London: Routledge, 2017), 229-41.

a single practice or training for a procedure called "copying". Neither has there been a single vocation or form of work that might be reducible to that of "copyist"'.[21] Jeroen de Kloet and Chow Yiu Fai also question the dominant framing of *shanzhai* as the pure production of the fake and the copy and suggest that the notion of *shanzhai* is 'less productive in explaining what is actually occurring in Dafen'.[22] Instead, they argue for more research on three aspects of *shanzhai* culture, namely, 'the aesthetics of the artworks or objects being produced; their local, national, and global circulation; and the aspiration of the people who are making *shanzhai* art'.[23] In order to gain more in-depth understanding of the discussion of 'the people who are making *shanzhai* art', I would like to go back to an often-quoted case in *shanzhai* studies: Leung Mee-ping's 'Made in Hong Kong/Shenzhen'.[24]

In 2006, Leung Mee-ping, a Hong Kong artist who was active in Hong Kong, Europe, and the United States, decided to temporarily live in Shenzhen. By disguising her identity as a practicing artist and by remaining anonymous, she posed as an amateur painter and went to Dafen Oil Painting Village to learn the craft of making replicas. After joining an artist's studio in Dafen Oil Painting Village and becoming an apprentice, Leung soon learned the regimented skills required to make replica paintings. Later, she began to place orders with the studio and requested replications of a number of photos that she had taken. The photos were all of popular tourist spots in Hong Kong, much-loved by mainland Chinese tourists. Some of those photos were taken in a clichéd and 'tacky' way, deliberately employing below-standard skills. The objects in the photos, which included Mickey Mouse pointing to Disneyland and horse racing in Happy Valley, were obviously chosen for tourists. Some subjects were so trivial that it was a surprise seeing them used as themes for the photos, for example the crowd outside a SaSa cosmetics store, the 'golden toilet' once owned by a Hong Kong tycoon, etc. All the paintings created from the photos, although not showing much (normative) artistic sense, became fundamental elements for the 'Made in Hong Kong/ Shenzhen' series. According to the artist, the project 'Made in Hong Kong/ Shenzhen', which included the contributions of the artisans in Dafen Oil Painting Village, created more than 400 artworks. With themes chosen from major tourist spots as well as typical images from Hong Kong movies – such as the starlight view of Victoria Harbour, Golden Bauhinia Square, Bruce

21 Wong, *Van Gogh on Demand*, 86.

22 De Kloet and Chow, '*Shanzhai* culture', 229.

23 Ibid.

24 Pang, *Creativity and Its Discontents*; Wong, *Van Gogh on Demand*.

Lee films and 'Young and Dangerous' – numerous copies of the paintings were reproduced for purchase.

In her book *Creativity and Its Discontents*, a study about creative industries and intellectual property in China, Lai-kwan Pang defined 'Made in Hong Kong/Shenzhen' as being an example of 'appropriation art', which uses 'massive replication' as the production mode to blur the existing and publicly recognised art line and also extended the discussion to authorship.[25] In other words, 'Made in Hong Kong/Shenzhen' as both art form and cultural representation is thereby seen as a carrier of critique in which various issues about the production and consumption of art, and the cultural dynamics between Hong Kong and mainlanders, can be explored. Whereas Pang's argument concerns the dynamics between the discourse of the global creative economy and Dafen as *shanzhai* culture, her conclusion is guided by artworks created by artists, but not, as de Kloet and Chow point out, *shanzhai* practitioners.[26] Winnie Wong's book-length study on Dafen, *VanGogh on Demand* (2013), provides another perspective in understanding Leung's work. Situating Dafen in the Chinese tradition of 'trade paintings', Wong makes reference to Leung's work when examining the discourses of originality, authenticity, creativity, and also the concept of 'speed' in the practices in Dafen.[27] As such, Wong's study opens up ways to rethink the relationships between creativity, craftsmanship, and *shanzhai* practices in Dafen.

Let us begin with the question: is 'creativity' a transcendental concept? In other words, can 'creativity' be detached from the contexts of cultural and social history and determine its own existence? Runco and Albert once provided a historical account for the concept of 'creativity'.[28] They argue that before Christianity dominated Western civilisation, creativity involved 'mystical powers of protection and good fortune'.[29] During the ancient Greek era, human beings' emphasis on the 'individual's daimon' turned genius into a secular thing. It was also when people began to see creativity as part of a person's own ability. In Roman civilisation, creativity became linked with the concept of 'genius', which was coupled with male dominance and the ability to have offspring. By saying that creativity is God's revelation in the

25 Pang, *Creativity and Its Discontents*, 217, 225-26.

26 De Kloet and Chow, '*Shanzhai* culture', 234.

27 Wong, *Van Gogh on Demand*, 93.

28 Mark Andrew Runco and Robert S. Albert, *Theories of Creativity* (New York: SAGE Publication, 1990).

29 Runco and Albert, 'A History of Research on Creativity', in *Handbook of Creativity*, ed. Robert S. Albert (Cambridge: Cambridge University Press, 1999), 18.

human world, the Bible has profoundly influenced Western culture since ancient times. During the eighteenth century, the Age of Enlightenment further added rationality to the concept of 'creativity' and linked it with 'personalisation'. At that time, unlike the previous belief that creativity is a kind of fate or God's blessing, it was recognised as being a person's individual ability. As Runco and Albert (1999) stated, 'The most significant distinctions made in the mid-1700s have to be the separations of idea of "creativity" from "genius", "originality", "talent" and "formal education". Since then, creativity is closely associated with "individual imagination"'.[30]

However, the discussion on creativity here is not only about its individualistic 'nature', but also about the culture of creativity that understands 'craftsmanship' as inferior in the creation of art. In 'Made in Hong Kong/Shenzhen', the artistic value of creativity focuses on the artist Leung Mee-ping, rather than turning her into one of the many painters at Dafen Oil Painting Village. In the conceptual framework that uses the individualistic model of creativity as the means to understand art, the entire set of over 400 paintings was 'done' by the 'only' artist in the production process, namely Leung Mee-ping. As such, the problematic mode and logic of artistic labour embedded in creative production is exposed. The 'problem' here is not about class exploitation, but the fact that it tends to overlook the role and status of craft labour, which is an integral part of creative production. In other words, both Pang's argument, which sees Leung's work as appropriation art and a way of understanding artwork in Dafen as *shanzhai*, can only be valid if we accept that craft labour is subordinate to artistic labour.

The discussion of craft labour is not new to the field of cultural studies.[31] Some studies, for example, have highlighted the need for using craft as 'workshop' in the creative market industries and the associated autonomy enjoyed by the practitioners.[32] Even though, in recent years, the necessity of craft labour in the creative industry and the potential decline of future craft production have been identified, the idea that craft labour is subordinate to

30 Runco and Albert, 'A History of Research on Creativity', 8.

31 See Glenn Adamson, *Thinking Through Craft* (Oxford: Berg, 2007); Peter Dormer, 'The Salon De Refuse?', in *The Culture of Craft*, ed. Peter Dormer (Manchester: Manchester University Press, 1997); Paul Greenhalgh, 'The History of Craft', in *The Culture of Craft*, ed. Peter Dormer (Manchester: Manchester University Press, 1997); Richard Sennett, *The Craftsman* (London: Allen Lane, 2007); Christina Hughes, 'Gender, Craft Labour and the Creative Sector', *International Journal of Cultural Policy* 18.4 (2012): 439-54; Mark Banks, 'Craft Labour and Creative Industries', *International Journal of Cultural Policy* 16.3 (8/2010): 305-21.

32 Banks, *The Politics of Cultural Work* (Basingstoke: Palgrave, 2007); 'Craft Labour and Creative Industries'.

artistic labour is persistent. Rewards for the hard work of the craft workers are often confined to their salary. They are frequently excluded from the prestige class and tend not to enjoy any fame. Craft workers are often unseen and unnoticed workers, frequently neglected by the public. For changes to take place, rational management and detailed division of labour are needed.[33] Today, the industrial and policy discourse about the creative industry is rooted in this individualised model of creativity, which tends to privilege artistic labour. From here, we can see a utopianised creative industry, in which these workers are considered to be creative and advanced. Unfortunately, we are not aware of the existence of other forms of creativity and labour, namely, craft labour, carried out by skill people who are able to perform delicate and meticulous work, and who are able to make artwork and creative products. As Andrew Ross has pointed out, rather than focusing on creative labour or high-tech skilled workers' livelihoods, scholars studying theories for creative industries should spend more time analysing those workers who do not get paid as original authors.[34] This point of view, I contend, is highly relevant in understanding the creative labour involved in art production in Dafen, especially in those cases where conceptual artists collaborate with Dafen artisans.

In an interview with Leung, I asked why only Leung's signature was on the paintings when many of the works were actually drawn by the painter-master who Leung collaborated with and his student-apprentices in Dafen Oil Painting Village. While assuming that the Dafen artists were not asked to sign, I was surprised by Leung's reply that it was actually the artists who refused to sign the work. When I asked for further explanation of this point, Leung said:

> Because they really don't care whether their signatures are on the works or whether they own their authorships of the works.[35]

The data I collected from my fieldwork in Dafen Oil Painting Village, however, suggest that this is debatable: it was not the artists in Dafen who did not care about authorship and signature; indeed, the Dafen artists do sign their 'original paintings'. The question is: what kinds of works are referred

33 Banks, 'Craft Labour and Creative Industries'.

34 Andrew Ross, *No-collar: The Humane Workplace and Its Hidden Costs* (New York: Basic Books, 2003); *Nice Work if You Can Get it: Life and Labour in Precarious Times* (New York: NY University Press, 2009).

35 Mee-ping Leung, personal interview, April 7, 2016.

to as 'original paintings'? In fact, the Dafen painters generally believe that works can be given authorship only when they are 'original paintings', i.e. when they are allowed to paint whatever they want. In such cases, even when they are not paintings that imitate Western art, but are tailor-made artworks made from customers' photos (i.e. logically, not a replica but an original), the artists generally do not regard these as 'original works' and do not sign the work (unless the customer requests it). In other words, non-original works are viewed as the sale and betrayal of craftsmanship and are not worth mentioning and definitely not 'good' enough to be signed. However, is a piece of art only a 'real' artwork when an author can determine 'what to draw'?

A story that Leung shared with me leads to a deep reflection on the politics of artistic and craft labour in artistic and creative production: On the day after Leung finished one of her series for the 'Made in Hong Kong/Shenzhen' exhibition, she received a handwritten letter from the painter-master who she had collaborated with for the Made in Hong Kong/Shenzhen project at the Dafen Oil Painting Village in Shenzhen. In the letter, the painter-master told Leung that he and his daughter happened to visit an exhibition one day, in which there was a set of works on display from the 'Made in Hong Kong/Shenzhen' series that had, in fact, been painted by the painter-master himself. When the painter-master and his daughter saw Leung's 'works', he felt a strong sense of alienation from 'his' painting. The 'revelation', or reunion with his work, shocked the painter-master and had an unexpected impact. His painting, without Leung's further technical input, had been the work of Leung which was exhibited at a gallery that he and his daughter visited. Rethinking his artistic identity and the value of his painting, the painter-master decided to close down his studio and workshop in the Dafen Oil Painting Village and to never accept an order for paintings or to collaborate with Leung again. Instead, he wanted to enrol at a postgraduate art school in China.

This is not a story about how Leung's conceptual art project in Dafen was able to empower the painter-master's artistic consciousness and identity. Instead, the story unveils the problematic power relations between artistic labour and craft labour; or, in McGuigan's terms, 'the architect and the bee',[36] in artistic production. In the contemporary art production paradigm, the individualistic model of creativity has been rooted and has overwhelmingly praised artistic labour and overlooked craft labour, not to mention '*shanzhai*

36 Jim McGuigan, 'Creative Labour, Cultural Work and Individualisation', *International Journal of Cultural Policy* 16.3 (8/2010): 323-35.

labour', made by those who have engaged in works without proposing 'creative' ideas, though they do employ 'creative' techniques and production skills. As such, the authorship and authenticity of craft labour have been neglected. Craft labour is not necessarily acknowledged as artistic, but the craft and creative contribution of its makers to the artwork should be rightly recognised. With an awareness of the value of craft labour, it is hard to address artistic production in Dafen as *shanzhai* art, and Leung's work 'Made in Hong Kong/Shenzhen' as appropriation art, because painters in Dafen, instead of being copyists, are essentially conducting craft labour, which, as Bank has stated, is 'an integral element of the creative industry labour process'.[37]

Conclusion

At this point, should *shanzhai* be referred to as a concept that is specific to China? At the beginning of April 2016, a portrait named 'The Next Rembrandt' was unveiled in Amsterdam, the Netherlands. As the name suggests, 'The Next Rembrandt'[38] is reminiscent of the Dutch master Rembrandt and is a 'new painting' made through contemporary 're-creation' of an original work by Rembrandt many years after his death. By using data and face identification technology to analyse Rembrandt's 346 portraits, art historians and technology experts created a large database. By using the system's facial recognition technology to identify and summarise the most classic geometric patterns of Rembrandt's works, a new painting was created by applying Rembrandt's skills and elements of his portrait features and style. The new portrait is of a 30- 40-year-old Caucasian man with a bearded face and wearing black clothes. If the 'The Next Rembrandt' project is regarded as a technologically advanced innovative art project created in a country known for its many famous artists, why has *shanzhai* art in Dafen not been endowed with a similar reputation?

One of the shortcomings in this study is that despite the fact that some artists at work were interviewed during the fieldwork in Dafen Oil Painting Village, I was unable to get in touch with the artists who were involved in the production of 'Made in Hong Kong/Shenzhen' to obtain first-hand

37 Banks, 'Craft Labour and Creative Industries', 306.

38 See original website of The Next Rembrandt. ING, *The Next Rembrandt Project*, A collaboration between ING, Microsoft, Tu Delft and Mauritshuis, assessed October 1, 2016, https://www.nextrembrandt.com.

information. According to the artist Leung Mee-ping, she has lost contact with both the aforementioned painter-master and the painters in his studio. During the interviews conducted at Dafen Oil Painting Village, it was impossible to contact any of the artists.

However, this also shows how mobile the painters in Dafen Oil Painting Village are. Many of the informants indicated that their painting skills are, in fact, no greater than their other job skills, and most of them would happily switch their jobs if they encountered job opportunities in other industries. These findings differ greatly from my those obtained by other studies of artists involved in art creation as artistic labour who call themselves artists or art practitioners. Although many of these artists did not regard art as their career, they would still choose various ways of living in order to be able to continuously accumulate cultural and social capital as artistic/craft labour and maintain their identity as artists and/or art practitioners.

Studying Leung Mee-ping's work 'Made in Hong Kong/Shenzhen', this chapter reveals the problematic divergence between artistic labour and craft labour, which frames the dominant understanding of and the creative labour involved in the concept of *shanzhai* art. Although we should be aware of the danger of celebrating *shanzhai*,[39] the contribution of craft labour made by, in this case, Dafen painters, to various forms of creative production should not be overlooked. The problematic power relation in which craft labour is subordinate to artistic labour is rooted in the individualistic model of creativity, which, I contend, does not recognise the creative input in various forms of mass production art. Although we see that the utopianisation of artistic labour as inherently creative and progressive has been diminished in recent years, there is still a failure to recognise the creative value of craft labour. *Shanzhai* studies, in this case, forces us to re-examine the current problematic understanding of authorship, creativity, and art.

Bibliography

Adamson, Glenn. *Thinking Through Craft* (Oxford: Berg, 2007).

Al Jazeera. 'Demand Grows for Chinese Fake Art – Al Jazeera English'. 2010. Accessed October 1, 2016. http://www.aljazeera.com/video/asia-pacific/2010/08/201083055011812822.html.

Banks, Mark. 'Craft Labour and Creative Industries', *International Journal of Cultural Policy* 16.3 (8/2010): 305-21.

39 De Kloet and Chow, '*Shanzhai* Culture', 233.

—. *The Politics of Cultural Work* (Basingstoke: Palgrave, 2007).

Cheung, Chi-fai. 'Planners Throw Out Arts Hub Objections Board Confirms Zoning Blueprint Giving West Kowloon Developer a Free Hand', *South China Morning Post*, December 13, 2003, p.1.

Cheung, Wai-sum. '冷靜與熱情之間：訪香港藝術家梁美萍，也談「展銷場：中國製造」', *Stand News*, June 23, 2015. Assessed October 1, 2016. https://thestandnews.com/art/冷靜與熱情之間-訪香港藝術家梁美萍-也談-展銷場-中國製造/.

China Daily. 'Shenzhen Shows Vitality, Dream with "Paintbrush" at Expo', August 26, 2010. Assessed October 1, 2016. http://www.chinadaily.com.cn/bizchina/shenzhen30years/2010-08/26/content_11209299.htm.

de Kloet, Jeroen, and Chow Yiu Fai. '*Shanzhai* Culture, Dafen Art, and Copyright'. In *Routledge Handbook of east Asian Popular Culture,* eds. Koichi Iwabuchi, Eva Stai and Chris Berry (London: Routledge, 2017), 229-41.

Dormer, Peter. 'The Salon De Refuse?'. In *The Culture of Craft*, ed. Peter Dormer (Manchester: Manchester University Press, 1997).

Greenhalgh, Paul. 'The History of Craft'. In *The Culture of Craft*, ed. Peter Dormer (Manchester: Manchester University Press, 1997).

McGuigan, Jim. 'Creative Labour, Cultural Work and Individualisation', *International Journal of Cultural Policy* 16.3 (8/2010): 323-35.

Hardt, Michael, and Antonio Negri. *Empire* (Cambridge, MA: Harvard University Press, 2000).

Headline Daily. '藝術去留系列 藝術創作 展示香港另一面 重新發掘本土', November 27, 2015. Assessed October 1, 2016. http://news.stheadline.com/dailynews/content_hk/2015/11/27/352899.asp.

Her Fatal Ways (表姐，妳好嘢！). Directed and written by Alfred Kin Ting Cheung, produced by Pooi-wa Chan (Hong Kong: Hong Kong: Golden Harvest, 1990). Assessed October 1, 2016. http://hkmdb.com/db/movies/view.mhtml?id=7280&display_set=eng.

Hughes, Christina. 'Gender, Craft Labour and the Creative Sector', *International Journal of Cultural Policy* 18.4 (2012): 439-54.

ING. *The Next Rembrandt Project*. A Collaboration between ING, Microsoft, Tu Delft and Mauritshuis. Assessed October 1, 2016. https://www.nextrembrandt.com.

Lai, Ming-hoi, and Eva Kit Wah Man. 與香港藝術對話：*1980-2014* (HK: Joint Publishing, 2015).

Lash, Scott, and Lury, Celia. *Global Culture industry: The Mediation of Things* (Cambridge: Polity, 2007).

Legislative Council of HKSAR. *Panel on Home Affairs and Panel on Development Joint Subcommittee to Monitor the Implementation of the West Kowloon Cultural District Project: The approved Development plan of the west Kowloon cultural district*) [LC

Paper No. CB(2)658/12-13(01)] (HK: HKSAR, 2013a). Assessed October 1, 2016. http://www.legco.gov.hk/yr12-13/english/panels/wkcd/papers/wkcd0225cb2-658-1-e.pdf.
—. *The Study of the West Kowloon Cultural District Development Project* (HK: HKSAR, 2013b). Assessed October 1, 2016. http://www.legco.gov.hk/yr1213/english/panels/wkcd/papers/wkcd0225cb2-658-1-e.pdf.
—. Legislative Council Subcommittee on West Kowloon Cultural District Development. *M+ in West Kowloon Cultural District* (HK: HKSAR, 2008). Assessed October 1, 2016. http://www.legco.gov.hk/yr04-05/english/hc/sub_com/hs02/papers/hs020524wkcd-546-e.pdf.
Leung, Mee-ping. Personal interview. April 7, 2016.
Leung, Sin-yue. '「被」遺棄的記憶——專訪香港藝術家梁美萍', *All about Macau Media* [論盡媒體], April 24, 2014. Assessed October 1 2016. http://www.inmediahk.net/node/1022480.
Li, Si-ming, Hung-ha Cheng, and Jun Wang. 'Making a Cultural Cluster in China: A Study of Dafen Oil Painting Village, Shenzhen', *Habitat International* 41 (1/2014): 156-64.
Li, Vivian. 'Original Imitations for Sale: Dafen and Artistic Commodification', *Art History* 37.4 (9/2014).
Lui, Tai-luk. 'City-branding Without Content: Hong Kong's Aborted West Kowloon Mega-project, 1998-2006', *International Development Planning Review* 0.3 (2008): 215-26.
Man, Eva Kit Wah. 'Influence of Global Aesthetics on Chinese Aesthetics: The Adaptation of Moxie and the Case of Dafen Cun', *Contemporary Aesthetics* 11 (2013).
Ng, Joyce. 'Arts Hub Seeks More Cash for Basement CEO of West Kowloon Cultural District Authority Says It Needs Extra Government Money in Coming Months to Build a Vast Underground Complex', *South China Morning Post*, July 26, 2012, p. 2.
Pang, Lai-kwan. *Creativity and Its Discontents: China's Creative Industries and Intellectual Property Rights Offenses* (Durham, NC; London: Duke University Press, 2012).
Ross, Andrew. *Nice Work if You can Get it: Life and Labour in Precarious Times* (New York: NY University Press, 2009).
—. *No-collar: The Humane Workplace and Its Hidden Costs* (New York: Basic Books, 2003).
Runco, Mark Andrew, and Robert S. Albert. 'A History of Research on Creativity'. In *Handbook of Creativity*, ed. Robert S. Albert (England: Cambridge University Press, 1999), 8-18.
—. *Theories of Creativity* (New York: SAGE Publication, 1990).
Sennett, Richard. *The Craftsman* (London: Allen Lane, 2007).

South China Morning Post. '1998 Former Chief Executive Tung Chee-hwa Announces Plan to Develop the West Kowloon Cultural District', September 30, 2011, p. 3.
Wen, Youping. *The Rise of Dafen Village* (China: Shenzhen: Haitian Press, 2007).
West Kowloon Cultural District Authority. Assessed October 1, 2016. http://www.westkowloon.hk/.
Wong, Winnie Won-yin. *Van Gogh on Demand: China and the Readymade* (Chicago, IL; London: University of Chicago Press, 2013).
Yang, Fan. *Faked in China: Nation Branding, Counterfeit Culture, and Globalization* (Bloomington, IN: Indiana University Press, 2015).
—. 'China's Fake Apple Store: Branded Space, Intellectual Property and the Global Culture Industry', *Theory, Culture and Society* 07 (2014).
Zhang, Lin, and Anthony Fung. 'The Myth of "*Shanzhai*" Culture and the Paradox of Digital Democracy in China', *Inter-Asia Cultural Studies* 14.3 (2013): 401-16.

About the Author

Louis Ho is Senior Lecturer of Humanities and Creative Writing at Hong Kong Baptist University. His research interests include cultural policy studies, cultural and creative industries, art and social innovation, museum and museology, and popular culture. Beyond the university, he has continued to participate in various arts and cultural organisations by serving as vice-chairman of 1a Space, Hong Kong Arts Development Council-3-year-granted art organisation, and committee vice-chairman of Renaissance Foundation Hong Kong, and committee member of the House of Hong Kong Literature.

2.3 Maoism and Disruptive Creativity

Shanzhai – an Alternative Perspective

Xiao Yuefan

Abstract

Shanzhai has become an umbrella term denoting the creative mimicry, parody, and counterfeiting that is pervasive in China. Yet, little has been agreed in terms of what has informed and underpinned the disruptive creativity that distinguishes *shanzhai* from exact copying. In this chapter, the author seeks to establish a conceptual framework for understanding *shanzhai* by drawing on Maoism. It is the author's contention that Maoism – especially its three founding pillars – art serves the people, the Mass Line, and self-reliance – has encouraged today's *Shanzhai* makers to challenge existing patterns of dominance and monopoly. The chapter concludes by suggesting that *shanzhai* will continue to play an important role in 'made/created in China' as long as Maoism is enshrined as part of the CCP's theoretical base.

Keywords: *shanzhai*, Maoism, Copyrights, The Mass Line, Great Leap Forward, self-reliance

Introduction

Chinese people have long been labelled 'uncreative'. Various empirical studies in the past have concluded that creativity was something discouraged by Confucian or collectivist Chinese culture.[1] However, the emergence of the *shanzhai* culture since 2008 has rendered such assertion debatable. Not

1 Steve M. Farmer, Pamela Tierney, and Kate Kung-McIntyre, 'Employee Creativity in Taiwan: An Application of Role Identity Theory', *The Academy of Management Journal* 46.5 (2003): 618-30; Elizabeth Rudowicz and Xiao-dong Yue, 'Compatibility of Chinese and Creative Personalities', *Creativity Research Journal* 14.3-4 (2002): 387-94.

Kloet, Jeroen de, Chow Yiu Fai, and Lena Scheen (eds), *Boredom,* Shanzhai, *and Digitisation in the Time of Creative China*. Taylor & Francis Group, 2019
DOI: 10.5117/9789462984745_CH2.3

exact reproducing but copying with ingenuity, *shanzhai* has become an umbrella term denoting creative fakery, mimicry, and modification pervasive in Chinese manufacturing, design, and art. For example, the GooApple handset – a *shanzhai* smartphone whose exterior looks almost identical to Apple's iPhone but which boasts an Android (i.e. Google) mobile operating system that has been customised to resemble the user interface of iOS (Apple's operating system for the iPhone). It suggests that the Chinese people are more than capable of merely imitating. Instead, their creations reveal that they are not shy in comfortably traversing conventionally defined divisions between systems of competition or opposition. Such traits expose Chinese creators' ability and willingness to innovate and transcend rigidity, for which the dominant understanding of Confucian traditions would suggest otherwise.

Moreover, 'creative industries' based on these creative practices have not only become an integral part of the Chinese (in)formal economy, but the resulting products have also made their way into the international economic arena and cultural sphere and, consequently, have the potential to destabilise the international use value and copyright regimes. As we have entered an era of globalisation where culture flows are becoming more mutual and thus the boundary between originators and receivers of culture are blurred,[2] we are more interested than ever in what *shanzhai* will mean for us. Is *shanzhai*, as Abbas suggests, a faking phenomenon that only proliferates at an early stage of globalisation?[3] To what extent is *shanzhai* not a natural outcome of cultural globalisation but a concerted effort at soft power promotion by the Chinese state? How is submission, adaptation, and resistance to authoritarianism and neo-liberalism in contemporary China epitomised by *shanzhai*? None of these questions can be answered without comprehending the origin and characteristics of this vernacularly Chinese notion that has travelled globally.

Understanding *Shanzhai*

To make sense of *shanzhai* is challenging. On the one hand, our understanding is bound by a convenient fake/real dichotomy, depending on which authority or legitimacy is attributed to hierarchically – a rigidity that the

2 Diana Crane, 'Culture and Globalization: Theoretical Models and Emerging Trends', in *Global Culture: Media, Arts, Policy, and Globalization*, eds. Diana Crane, Nobuko Kawashima, and Ken'ichi Kawasaki (London: Routledge, 2002), 1-28.

3 Ackbar Abbas, 'Faking Globalization', in *Other Cities, Other Worlds: Urban Imaginaries in a Globalizing Age*, ed. Andreas Huyssen (New York: Duke University Press, 2008), 243-66.

conceptualisation of *shanzhai* in terms of intellectual property rights has contributed to in no small way.[4] As de Kloet and Scheen point out, we should be cautious about simplistically dismissing the alleged copy and validating the assumed original when interpreting *shanzhai*. Instead, *shanzhai* should be read as a process of translation that enables more imaginations and iterations.[5] On the other hand, *shanzhai* is an overarching term encompassing a variety of creative practices in the economic, social, and aesthetical realms, the speciality of each case making it even harder to make generalisations. Nevertheless, scholarly attempts have been made to tap into this cultural consciousness so rich in expressions. The most popular form of enquiry has been the use of case studies to illustrate what *shanzhai* stands for vis-à-vis state, society, and modernity. Chubb focuses on the production and consumption of a range of mundane, joyfully pirated products and regards the *shanzhai* phenomenon as a grass-roots reinterpretation of dominance that worships as much as it mocks authority.[6] De Kloet and Scheen use the example of Pudong and argue that urban experiments there are most emblematic of a *shanzhai* imaginary global city.[7] Pudong is neither generic, nor fake since it is characterised by a distinctive spatial arrangement (vis-à-vis the Bund); it inherits its own colonial and communist history, and is replete with authentic local experience. In this sense, *shanzhai* neutralises the homogenisation impulse of globalisation.

For *shanzhai* art, Wong presents her ethnographical research in Dafen Oil Painting Village, famous for its mass production of copy masterpieces, and argues that 'replica' can no longer capture the creativeness of artworks produced there as they exhibit considerable traits of 'transfer, transformation, invention, innovation, appropriation and delegation'.[8] On performance, Zhang and Fung investigate the *shanzhai* Spring Festival Gala – an internet-based parody, staged by the marginalzed, of the state-run Gala.[9] They trace the trajectory of this supposed manifestation of resistance, from a grass-roots amateur initiative, through commercialisation by the digital economy and co-optation by local state media, to suppression by the central government.

4 Fan Yang, 'China's "Fake" Apple Store: Branded Space, Intellectual Property and the Global Culture Industry', *Theory, Culture & Society* 31.4 (2014): 71-96.

5 Jeroen de Kloet and Lena Scheen, 'Pudong: The *Shanzhai* Global City', *European Journal of Cultural Studies* 16.6 (2013): 692-709.

6 Andrew Chubb, 'China's *Shanzhai* Culture: "Grabism" and the Politics of Hybridity', *Journal of Contemporary China* 24.92 (2015): 260-79.

7 De Kloet and Scheen, 'Pudong'.

8 Winnie Wong, *Van Gogh on Demand: China and the Readymade* (Chicago, IL; London: University of Chicago Press, 2014), 87.

9 Lin Zhang and Anthony Fung, 'The Myth of "*Shanzhai*" Culture and the Paradox of Digital Democracy in China', *Inter-Asia Cultural Studies* 14.3 (2013): 401-16.

Their study reveals the limits that allow digital technology to garner, channel, and materialise popular imaginary for political democracy in China. The democratising aspiration bestowed on *shanzhai* is thus overstated as it represents little more than 'a site of mediation'.[10] *Shanzhai* reproduces, if not reinforces, existing patterns of predominance by state capitalism.

Overall, *shanzhai* is a grab-bag notion with confounding implications, evident in the above studies. Despite uncovering the defining characteristics of *shanzhai* via different case studies, these studies do share an understanding of the etymology of *shanzhai*. Exemplified by the painstaking work of Hennessey on deconstructing this notion, *shanzhai* is generally thought to originate from Chinese historical romances featuring rebellious tribes fighting for justice from remote, mountainous (*shan* 山) strongholds (*zhai* 寨) beyond the control of imperial courts, epitomised by the novel *Shuihu Zhuan* (水浒传).[11] In this novel, Robin Hood-like legends flee to the mountainous stronghold (*shanzhai*山寨) to evade capture or punishment by corrupt officials.[12] They fight ruthlessly with government troops commanded by these officials while at the same time still holding a glimmer of hope of being recognised by the emperor – out of a paradoxical conviction that they are demonstrating their loyalty to the emperor by resisting those evil local tyrants. It is this moral dilemma of these ancient righteous warriors that still informs the ambiguity embedded in the *shanzhai* phenomenon today.

Moreover, for the vast majority of authors researching *shanzhai* as creative practices, it is a rather recent phenomenon. *Shanzhai* activities are said to be more visible and rampant in more globalised parts of China. In particular, the Special Economic Zone of Shenzhen is regarded as the home of *shanzhai* and China's copy capital.[13] In most cases, it looks as if *shanzhai* creative activities only started to surface *after* China was integrated into the global political economy.

Interestingly, in most works, there seems to be a *chuanyue* (穿越) – a vernacular term in Chinese meaning that an ancient thing or person has resurged in the contemporary context – in the historical evolution of *shanzhai*. In other words, for most authors, *shanzhai* has been posited as a lost ethos not reactivated until China goes global. From the *Shuihu Zhuan* era of the twelfth century to the new millennium, the surprising 'leap' in

10 Ibid., 412.

11 William Hennessey, 'Deconstructing *Shanzhai* – China's Copycat Counterculture: Catch Me If You Can', *Campbell Law Review* 34.3 (2012): 609-60.

12 Yang, *Faked in China* (Bloomington, IN: Indiana University Press, 2015).

13 Michael Keane, *China's New Creative Clusters: Governance, Human Capital and Investment* (London: Routledge, 2013a).

the analysis of *shanzhai's* historical lineage is where the existing literature falls short. To be sure, the analyses by Hennessey and Chubb have tried to address this problem. Hennessey extends the time span to cover the entire imperial China period and suggests that the counterculture of *shanzhai* continued into the neo-Confucian era, whereas the culture of imitation was endorsed and favoured by the establishment.[14] Chubb, by contrast, narrows this gap further by drawing a close link between *shanzhai* and Grabism (*nalai zhuyi,* 拿来主义) proposed by Lu Xun in 1934.[15] Grabism is a pragmatic approach emphasising that no matter the origins of useful things, be they foreign or indigenous, they should be borrowed, inherited, or mixed up as long as the new combination works for China's modernisation.

Despite this, the decades from the 1930s until the late 1970s – a period of profound social experiment, transformation, and upheaval – are still missing from existing analyses. This period also overlaps with the entire political career of Mao Zedong. The task of this article is to fill this critical gap in understanding what has informed *shanzhai*. There has been a serious enquiry into contemporary *shanzhai* as a playful phenomenon due in large part to the spirits of populist anarchism, irreverence, anti-establishmentism, illegality, and nationalism. The courts of imperial China and the present Chinese Communist Party (CCP) regime are equally vigilant to these spirits. However, there was a period when all these spirits were recognised not only as legitimate but as avant-garde. Indeed, in the Mao era these spirits and the chaotic social fluidity they fuelled were celebrated as progress.

In the English language literature, this Maoist dimension of *shanzhai* has remained undiscovered. However, this is not the case with the Chinese language literature. When performing a keywords search on the CNKI (China National Knowledge Infrastructure) database,[16] a sizeable number of articles draw close links between 'the Mass Line' (群众路线)[17] (148 articles), 'mass' favourite (群众喜闻乐见) (144 articles), and 'created by the masses' (群众发明创造) (42 articles), on the one hand, and *shanzhai* on the other. However, despite frequent references to Maoist rhetoric, they fail to advance and conduct a systematic analysis of the intrinsic linkage between different tenets of Maoism and *shanzhai's* expression in various forms.

14 Hennessey, 'Deconstructing *Shanzhai*'.

15 Andrew Chubb, 'China's *Shanzhai* Culture: "Grabism" and the Politics of Hybridity', *Journal of Contemporary China* 24.92 (2015): 260-79.

16 The database search as restricted to the publication period 2008-2015, since 2008 is widely regarded as *shanzhai*'s 'epic year'.

17 This communist revolutionary term refers to a leadership method that pays attention to consulting the masses, and transforming their opinions into actions.

Therefore, from a China Studies perspective, it is my contention that Maoism, too, has informed and underpinned the current *shanzhai* culture. Maoism, also known as Mao Zedong Thought treats the relationship between art and politics, the Mass Line, and self-reliance. These teachings have decisively pre-configured the contours within which present-day *shanzhai* activities operate. The following sections will proceed by tracing the roots of *shanzhai* in these three founding pillars of Mao Zedong Thought.

Art and Politics: To Rebel is Justified

Today, *shanzhai* is partly but not exclusively about anti-establishmentism. After all, *shanzhai* as a form of resistance against (neo)Confucianism existed throughout dynastic China, either in a violent *shanzhai*-style guerrilla warfare against corrupt imperial courts, or in other forms of disobedience such as the underground circulation of outlawed publications. What has enriched this rebellious mentality to become the spirit of contemporary *shanzhai*, however, is the liberal ideas of social equality and populism that enable commoners' active participation in social, cultural, and political life on a more equal footing. Since the inception of his political career, Mao Zedong was at the forefront of negotiating and adapting these liberal values to the Chinese context.

Like his fellow intellectuals in the iconoclastic New Culture Movement, Mao regarded Confucianism as China's social malaise. For him, the corrupt and rotten Confucian culture maintained a powerful system of exclusion that was responsible for a fatalistic and passive population.[18] Based on the Confucian moral code of *sangang* '三纲' (three bonds: the minister was subject to the guidance of the monarch, child to father, and wife to husband), imperial China was characterised by a hierarchical and oppressive social order wherein every member of the society was confined to his/her particular position as dictated by *sangang*. Failure to fulfil one's obligation as designated by his/her particular social role or attempts to break away from this system of exclusion were seen as treason and heresy. Consequently, imperial China achieved a state (illusion) of harmony by virtue of societal inertia. According to Han Feizi, on whose thoughts *sangang* was founded, 'when the three bonds are adhered to, there is good order; when the three bonds are breached, there is great disorder'.[19]

18 Kenneth Lieberthal, *Governing China: From Revolution Through Reform*. 2nd edition. (New York: W. W. Norton & Company, 2004), 61.

19 Cited in Zenghua Shao, 韓非子今註今譯 [*Contemporary annotation and interpretation of Han Fei Zi*] (Taipei: Commercial Press Taiwan, 1990), 1004.

There was an artistic dimension to Mao's critique of Confucianism. For him, 'feudal' art and literature played an important part in sustaining this system of exclusion. The ruling class used this art and literature as a means to instill a sense of perpetual obedience to the superior among the inferior. The promotion of hierarchical values was pervasive in literature, music, and painting.[20]

Soon after Mao realised that China's 1911 'bourgeois democratic' revolution had failed to materialise aspiration for national independence and cultural renewal, he continued to wage an attack on Chinese traditions and extended the targets, this time to bourgeois and petty bourgeois cultures. He argued that these 'new' cultures did little to emancipate or enlighten the poorly educated masses due to their inaccessibility in terms of language and genre. What most bourgeois writers and artists had been preoccupied with, he complained, were hollow, artistic creations, full of affectation and narcissism, which lost touch with both the harsh reality of China's 'semi-feudal and semi-colonial' inferior status and people's grievances.

For Mao, Confucian, bourgeois, and petty bourgeois cultures together formed an impregnable system of exclusion. By dictating and legitimising what can be said, heard, or seen in the art world and in everyday conduct, this system of exclusion barred the socially subordinate, economically disadvantaged, and the illiterate from speaking out and kept their aspirations hidden. Effectively, commoners and the proletariat played a negligible role in the cultural life of imperial and Republican China as a result of alienation, a reflection of their minimal political participation.

Ironically, the marginal status of the pre-PRC underclass resurfaced in the post-socialist PRC where state capitalism prevails and varieties of Confucianism are revived. This is also the social and cultural context in which present-day *shanzhai* has emerged. Isn't *shanzhai* a salient expression of the participatory aspiration of those excluded? In what follows, the author will delineate how Mao's lifelong struggle for more equality and social mobility (at least more dynamism in an inertia society) – through his fight with the Confucian, bourgeois, and Leninist systems of exclusion – informed today's *shanzhai* spirit.

Mao systematically elucidated the need to challenge the dominant system of exclusion through aesthetic intervention in his 1942 *Talks at the Yan'an Forum on Literature and Art*. Mao called for smashing the hierarchies in the aesthetic order. The *Talks*, and especially its emphasis on promoting

20 Qiaobin Deng and Zhongzhu Jiao, '封建社会的君主专制与文艺功能论 [On auto cratic monarchy and the function of literature and art in the feudal society]', 浙江大学学报(人文社会科学版) [*Journal of Zhejiang University* (*Humanities and Social Sciences*)] 33.1 (2003): 97-104.

art for and by the 'uncultured', was to become the source of legitimacy underpinning *shanzhai* art.

The *Talks* were meant to clarify two issues: who art was for and what role artists should play. As to the first question regarding audience and the class affiliation of art, Mao stated that the CCP believed that art was for 'the broadest sections of the people'.[21] Art should give a voice to those who had previously been excluded by the social order maintained by such exploiters and oppressors as the landlord class, the bourgeoisie, and the imperialists. Thus, artists are required to appeal to a much larger section of society and feature prominently in the lives and stories of the underclass.

With regard to the order of importance between art and politics, Mao was apathetic to the idea of 'art for art's sake' and argued for the subordination of art to politics and to (proletarian) class interests.[22] He thus saw a reconfiguration of the aesthetic arena as a mobilisation vehicle for oppositional politics and for social equality at large.

On the second issue of how to be a good artist, Mao's answer has had profound implications on creativity and *shanzhai*. He argued that artists' service to society comprised two tasks: to popularise art and to raise people's ability to appreciate and create art.[23] In order to fulfil these two roles, artists needed to wholeheartedly learn from the people what kind of language or genre made the most sense to them. 'Only by being their pupil can he [an artist] be their teacher'.[24] In particular, Mao valued nascent artistic creations by the masses, such as wall newspapers, murals, folk songs, and folk tales, and urged artists not to dismiss these as primitive but to help the masses improve the quality of their work.

This approach encouraged what Wong later found in her discussion of *shanzhai* art: the deskilling of art into mass culture.[25] More precisely, first, the definition of art has been extended considerably; nascent artistic creations by art amateurs from all walks of life are no longer considered unqualified. Second, *anyone*, with some training, can become an artist.

21 Mao Zedong, '在延安文艺座谈会上的讲话 [Talks at the Yan'an Forum on Literature and Art]', *News of the CPC*, May, 1942, accessed May 15, 2015, http://cpc.people.com.cn/GB/64184/64185/66617/4488965.html.

22 Mao, '在延安文艺座谈会上的讲话 [Talks at the Yan'an Forum on Literature and Art]'; Lai-kwan Pang, *The Art of Cloning: Creative Production during China's Cultural Revolution* (London: Verso, 2017).

23 Mao, '在延安文艺座谈会上的讲话 [Talks at the Yan'an Forum on Literature and Art]'.

24 Ibid.

25 Wong, *Van Gogh on Demand*.

In both cases, the aesthetic arena has been pluralised. This also provides present-day *shanzhai* with legitimacy.

Mao's efforts to reconfigure the aesthetic arena continued into the PRC era. He had been constantly vigilant on any reactionary comeback by former cultural elites. However, by the early 1960s, he realised that the vanguard Party had gradually become the vested interest and formed a new, oppressive system of exclusion supported by reactionary culture and Leninist means of social control. He decided to revive the ideas and ideals he had developed earlier in the iconoclastic New Culture Movement and in his 1942 *Talks* in a more radical way, i.e. the Great Proletarian Culture Revolution (CR).

Although it is beyond the scope of this article to analyse the trajectory of the CR, its aesthetic dimension warrants highlighting. That is, Mao mobilised the people to destroy the Party by granting them the Four Freedoms: 'to speak out freely; to air views fully; to hold great debates; and to write Big Character Posters (大鸣，大放，大辩论，大字报)'.[26] The masses not only had unprecedented freedom to publish almost unconstrained as Party organisations, public security apparatus, and wider social hierarchies were stormed, but once again emerged at the forefront of the aesthetic arena.[27] The Four Freedoms were also enshrined in the 1975 and 1978 versions of PRC constitution. It should be noted, though, that the briefly available freedom of expression as a result of anarchy can never explain away the unprecedented atrocities that occurred during the CR. The movement lost control soon after its inception and quickly became a project caught between emancipation and manipulation.

What does the CR have to do with *shanzhai*? Yu Hua characterises the CR as a period of *shanzhai* when 'the revolutionary inner nature of disadvantaged groups in society was activated' to overthrow the establishment under the Maoist rubric of 'to rebel is justified'.[28] He specifically points to anarchism during the CR, when myriad autonomous mass organisations were formed,[29] just like rebellious strongholds in ancient times. Moreover, the participatory struggle in the aesthetic arena by the masses during the CR serves as a beacon that often lends legitimacy to present-day *shanzhai* cultural practices. To be sure, the current underclass enjoys few of the Four Freedoms. They are, nevertheless, constantly contesting the dominant statism-neo-liberalism

26 Linda Benson, *China since 1949* (London; New York: Routledge, 2016), 189.

27 Pang, *The Art of Cloning*.

28 Yu Hua, 十个词汇里的中国 [*China in Ten Words*] (Taipei: Rye Field Publishing Co., 2011), 247.

29 Ibid.

nexus in the form of *shanzhai* art and performances that carry their aspirations to participate in contemporary social and political life.

The Mass Line: from the Masses, to the Masses

Before uncovering the historical linkage between *shanzhai* and the Mass Line, it is important to recognise the often neglected distinctions between Maoism, Marxism, and Leninism. For Mao, the latter two were often referred to as 'dogmas' – orthodoxy taken for granted without appreciating peculiar Chinese conditions.[30] Defying dogmas with grass-roots creativity derived from practice – another defining characteristic inherent in contemporary *Shanzhai* – first acquired its theoretical underpinning in Mao's criticism of Leninism for its lack of popular participation in policymaking. In contrast to Leninism, where the people were perceived as ignorant (thus denied access to policy debates), Mao's faith lay in the masses, whom he believed were the source of true wisdom.[31]

This preference was reflected in his long-term suspicion of intellectuals and the knowledge they possessed. Intellectuals were deeply distasteful to Mao for two reasons. One was that he regarded intellectuals as bookworms, who simply claimed expertise from their dogmatic studies of theories but knew little about how things work on the ground, as they were reluctant to 'dirty their hands by going out among the people to learn about real conditions'.[32] In particular, according to orthodox theories, intellectuals well-versed in Western theories (including returnees from the Soviet Union) often hastily asserted the technological infeasibility of achieving something great given China's inferior productivity. They did so before actually learning from workers and peasants who creatively developed quick fixes or shortcuts that, as Mao believed, overcame or at least 'got around' the same obstacles. In other words, unorthodox or *shanzhai* methods invented by the masses were valuable approaches that should be cherished because they could potentially help achieve the conventionally unthinkable, even though the finished products were barely usable. This later becomes a supporting argument often proudly cited by today's *shanzhai* makers, who claim better knowledge of realities on the shop floor and accordingly

30 Shaun Breslin, *Mao: Profiles in Power* (Harlow: Longman, 2000).

31 Ibid.

32 Lieberthal, *Governing China*, 71.

renovate (i.e. simplify) the production process, something that established manufacturers are unwilling to risk due to their conservative inclinations.[33]

The other reason was Mao's concern that the Confucian social order – in which the literati were afforded a higher status than the common people – would revive if the newly founded People's Republic were to rely too heavily on armchair experts, intellectuals, and managers for state building and economic construction.[34] For Mao, the political loyalty of this echelon was questionable. He believed that, despite the revolution, they still had 'attitudinal hangovers' about their superiority over the proletarian masses.[35] Instating intellectuals to leading roles would undo his efforts to achieve a more equal relationship between intellectual and manual labour. He went so far as to contend that:

> Scientific knowledge and technological inventions are usually created by classes under suppression. In other words, they are created by those with a low social status, little education, in poor [financial and living] conditions; by those whom at the beginning always being laughed at, or even attacked, tortured and executed.[36]

This interpretation of the true source of creativity, I contend, provided considerable justification and created an ambiguous safety zone for the development of contemporary *shanzhai*.

The masses, emblematic of manual labour, were thus not only worthy, but also trustworthy to listen to. When it came to the creative potential of the masses, Mao praised them for having '[...] boundless creative power. They can organise themselves [...] [and] concentrate on production in breadth and depth and create more and more undertakings for their own well-being'.[37] His voluntarist confidence in the masses – the belief in people's unconquerable volition to the extent that once they were fully mobilised in concentrated efforts they could overcome any challenges, regardless of the prevailing productivity level – set him further apart from orthodox Marxism in which

33 Bai Ye, '山寨文化纵横谈 [Open discussion on *shanzhai* culture]', 社会科学 论坛 (学术评论卷) [*Tribune of Social Sciences*] 2 (2009): 56-90.

34 Breslin, *Mao*.

35 Ibid., 59.

36 Mao, 1985, cited in Guisheng Chen, 教育研究空间的探求 [*Explore the space for education and research*] (Fujian Education Press, 2006), 308.

37 Mao, 'The Mass Line', in *Quotations from Mao Tse Tung* (Peking Foreign Languages Press, 1966), accessed May 11, 2015, https://www.marxists.org/reference/archive/mao/works/red-book/ch11.htm.

materialism is the incontestable cornerstone.[38] It was also heterodox in comparison with classic doctrines of socialist economic construction emphasising socialist rationality in the forms of central planning and top-down techno-bureaucratism – an orthodox approach to economic advancement preferred by Mao's Leninist-leaning colleagues but which Mao considered uncreative or even suppressive of mass motivation.

Mao also had a unique take on how the Party-state could best utilise available forces of production to lift China out of backwardness. He characterised this underdeveloped condition as 'poor and blank' (一穷二白), that is, poor as a result of a low level of industrialisation and blank in the sense that China had little accumulated knowledge on (Western- and Soviet-dominated) science and technology.[39] However, Mao regarded 'poor and blank' as an advantage,[40] because being a latecomer also meant that China could develop itself relatively unconstrained by the dogmatised experience of developed countries and create its own path for modernisation. In Mao's words:

> The poor want a revolution, whereas it is difficult for the rich to want a revolution. Countries with a high scientific and technological level are overblown with arrogance. We are like a blank sheet of paper, which is good for writing on.[41]

This interpretation of the advantage of backwardness foreshadowed Mao's subsequent appeal for launching a technological revolution by utilising mass creativity and mass political energy rather than relying on dogmatised knowledge. The pinnacle of this 'technological revolution' was what later became known as the Great Leap Forward (GLF) – which I consider a mass *shanzhai* campaign that most of China's working population at the time participated in – shed important light on contemporary *shanzhai*.

While Mao also acknowledged the importance of learning from abroad, albeit with an 'analytical and critical' eye for foreign knowledge and experience,[42] he believed this was insufficient to arouse the political enthusiasm of the proletarian masses. Mao wanted to instill a sense of

38 Lieberthal, *Governing China*; Maurice Meisner, *Mao Zedong: A Political and Intellectual Portrait* (Cambridge: Polity, 2007).

39 Mao, '论十大关系 [On the Ten Major Relationships]', *Xinhuanet*, December 30, 2004, 1956, accessed June 10, 2015, http://news.xinhuanet.com/ziliao/2004-12/30/content_2393996.htm.

40 Meisner, *Mao Zedong*.

41 Mao, '论十大关系 [On the Ten Major Relationships]'.

42 Ibid.

ownership, to the point that the people saw the Party's undertakings as their own.[43] He sought to cultivate an active and voluntarist subjectivity.

The hallmark of the governing mode that Mao endorsed thus comprised an accessible Party in listening mode and a dynamic populace, motivated to adopt initiatives not so much by coercion but by virtue of their spontaneous embracement of socialist citizenship. This engaging relationship formed the basis of the organising principle of socialist government – the Mass Line. Its mass orientation was reflected in the governing principle 'from the masses, to the masses'.[44] That is to say, cadres should elicit disparate ideas from the masses that could potentially improve policy outcomes. As a result, when these policies based on mass input were 'returned' to the masses, they would be welcomed by the masses themselves because of their prior involvement in the policy formation process. The Mass Line was deemed a virtuous loop in which the people were convinced of their political efficacy and, at the same time, of the economic progress, embodying Maoist governing strategy that combined economic development with socialist awareness.[45]

Furthermore, the applicability of 'from the masses, to the masses' was extended to all policy areas and was upheld as the *universal* technologies of government from the mid-Mao era onwards. As an editorial in the *People's Daily* pointed out on the eve of the GLF: 'the technological revolution [...] that the Party is calling for, just like the socialist revolution in economic, political and ideological affairs, must also be a mass movement' based on the premises of 'trusting mass power and reflecting mass desire'.[46] Henceforth, indigenous and creative (albeit primitive) techniques invented by the masses, or *shanzhai* methods, had shaken off all heterodox connotations. *Shanzhai* became *the* established doctrine of problem solving, whereas more scientific methods with a focus on technological sophistication, but which were disdainful of mass creativity, were marginalised.

As rationality gave way to passion and techno-bureaucratism to voluntarism, it was not surprising that the GLF unfolded in a *shanzhai* manner. This was clearly evident in the General Line (总路线) – the guiding principle of socialist construction: Go all out; aim high; build socialism rapidly and economically with better and greater results (鼓足干劲，力争上游，多快好省地建设社会主义). Given the promotion of mass creativity at the

43 Breslin, *Mao*.

44 Gary Sigley, 'Chinese Governmentalities: Government, Governance and the Socialist Market Economy', *Economy and Society* 35.4 (2006): 487-508.

45 Breslin, *Mao*.

46 *People's Daily*, '社论：把总路线的红旗插遍全国 [Editorial: Stick the Red Flag of the General Line All over the Country]', May 29, 1958a.

expense of scientific knowledge, unconstrained mass initiatives unleashed great destructive power that led to catastrophes. For instance, the Great Famine following the GLF was largely due to a voluntarist belief that 'The crop from the field is as large as guts can yield' (人有多大胆，地有多大产).

The most ironic episode of the GLF that directly mirrors contemporary *shanzhai* practices was utilising mass creativity and ingenuity for steel making. In 1958, Mao proposed an ultra ambitious goal of doubling steel output in the same year, with a view to eventually overtaking the UK and the US in steel production in a matter of a decade. This target was clearly unattainable according to orthodox theories of industrial development and prior experience of industrialised countries. Yet, for Mao, revolutionary fever among the masses would effectively compensate for the shortage of material resources. Immense mass initiative and creativity was a formidable drive of production in its own right. As such, he resorted to the Mass Line and favoured the *shanzhai* approach.

The masses in all factories, even in schools and hospitals, were accordingly mobilised to make steel. Having no knowledge of metallurgy and no experience in steel making, they relied upon a variety of indigenous but wildly primitive methods that they created 'out of the box'. To name a few of these *shanzhai* innovations: using gasoline cans as backyard steel furnaces; melting pots, pans, and other metal household articles as raw materials; using firewood as fuels.[47] Despite being given credit for their ingenuity, the nationwide implementation of these *shanzhai* methods produced vast quantities of useless pig iron.

In the economy as a whole, the race towards rapid industrialisation was also carried out in a *shanzhai* manner. The masses in all industrial sectors proclaimed and boasted that their *shanzhai* methods embodied 'greater, quicker, better and cheaper' (多快好省):

> In the current great development of technological revolution, [we] have been creating using primitive methods superior and state of the art products that have never existed before. These are all uniquely stylish products with simpler structures and greater efficiency.[48]

Examples of the above were numerous, mostly multifarious prototypes, such as automobiles, locomotives, and machinery – novel but barely usable and

47 *People's Daily*, '土法炼钢 妙计无穷 [An Unlimited Number of Ingenious Techniques That Were Indigenously Developed to Make Steel]', October 5, 1958b.

48 Mao, 1960, cited in Yunhe Zhu, '从"土洋并举"到"以土为主" – "大跃进"时期技术革命方法的嬗变 [From "combining indigenous and foreign techniques" to "mainly rely on indigenous techniques" – the evolution of methods deployed in technological revolution during the Great Leap Forward]', 北京党史 [*Beijing Party History*], 6 (2011): 28.

short-lived, too unreliable to adopt for scale production. Unsurprisingly, the wider GLF shared a similar fate as *shanzhai* steel making.

Despite little achievement in economic terms and a disastrous ending, the *shanzhai*-style GLF left a profound mark on a mentality of rationality that survived into the post-Mao era. It should be noted that the term *shanzhai* never appeared in official discourse, but it was precisely because of its newly acquired orthodox status that a more 'politically correct' alternative was used and, in this case, mass creativity instead of *shanzhai*. It would be rather simplistic to suggest that in the specific context of 1950s PRC these two terms were interchangeable. Nonetheless, *shanzhai* indeed closely resembled the Mass Line ethos for three reasons, such that when the former resurfaced decades later, the way it sprawled also retrospectively mirrored the latter. I will explain such parallels using the example of *shanzhai* mobile phones – *the* totem of everything *shanzhai*.[49]

First, the Mass Line ethos of 'greater, quicker, better and cheaper' was substantiated by tactics aimed at overcoming resource constraints. These could be epitomised by 'make the best use of local materials and conditions' (就地取材，因陋就简).[50] By extension, it meant substituting whatever materials were available locally for key materials or components that could not be made from local efforts; or to simplify complex processes and designs so that people with little education or experience would still be able to learn and make products in large quantities. These mass created tactics were the combination of a defiance of established rules and economic imperative. At best, these interim measures can be seen as tactics that economise production. At worst, they were tricks to cut corners. Present-day *shanzhai* practices are embedded with features of both.

Shanzhai products are famous for their value for money but seldom for durability. Take *shanzhai* mobile phones: As a result of being *shanzhai*, most of these phones are made outside of the formal economy, hence rarely follow national quality assurance standards or conventional manufacturing procedures adopted by established brands. It is common for *shanzhai* makers to operate without licences and *shanzhai* phones do not undergo a thorough quality check before shipment. This makes them very competitive in terms of cost and turnaround unmatched by established brands.[51] Simplified

49 Yang, *Faked in China*.

50 *People's Daily*, '上海的技术革命群众运动 曹荻秋代表的发言 [The Mass Campaign of Technological Revolution in Shanghai – Speech by Representative Cao Diqiu]', April 9, 1960.

51 Cara Wallis and Lin-chuan Qiu, '*Shanzhaiji* and the Transformation of the Local Mediascape in Shenzhen', in *Mapping Media in China: Region, Province, Locality*, eds. Wanning Sun and Jenny Chio (London; New York: Routledge, 2012), 109-25.

processes have also transformed mobile phone making from high-tech to low-tech, thus enabling amateur makers with little formal knowledge of telecommunications and design basics to enter the market. In this respect, *shanzhai* phone making has become more of a creative than a manufacturing process – little R&D is involved, all *shanzhai* makers need to do is to make their phones functional and aesthetically appealing. As to functions, *shanzhai* phones are marked by their versatility – 'all-in-one' designs combining several functional modules, e.g. a handset with a built-in TV tuner – are made available for the 'information have-less' at a fraction of the cost of a mainstream phone.[52] Another extreme example is the Daxian *shanzhai* phone, which achieves its boasted 'wireless' charging functionality by simply mounting a built-in power plug and voltage transformer on the handset.[53]

Admittedly, free market fundamentalism matters,[54] in that for the producers of these phones, *shanzhai* is a vital surviving strategy. Even without the influence of Maoism and the Mass Line, cutting corners and other economising tactics would still seem the optimal choice for *shanzhai* makers who compete in the market by undercutting mainstream producers. In other words, *shanzhai* making in China is not very different from underground manufacturing elsewhere. This raises the question of differentiating the economic versus the cultural dimensions of *shanzhai*, hence the possibility of *shanzhai* without democratic connotations but purely as an imperative for economic survival.[55] However, this is certainly not the whole story about Chinese *shanzhai* manufacturing in general, and mobile phone making in particular. Although the economic survival perspective is insightful with regard to why *shanzhai* makers produce *cheap* phones, it lacks explanatory strength when it comes to analysing *creativity* embedded in these phones. That is to say, creative elements and how they come about require analysis beyond neo-liberal market competition.

This is especially true if we take a closer look at those 'rebellious' traits, such as novel product design in contravention of established knowledge and the belief of *shanzhai* makers that manual labour can make products as great as intellectual labour (thus bringing down the hierarchy of labour in the creative domain). It is therefore this unique blending of economic survival strategy underpinned by neo-liberalism and democratising connotation

52 Ibid.

53 Youtube, '山寨手機夠晒堅 神奇無線充電 [Magical *Shanzhai* Phone with Wireless Charging Functionality]', 2013, accessed August 15, 2016, https://www.youtube.com/watch?v=eX--pvEWNso.

54 Wallis and Qiu, '*Shanzhai*'.

55 I'm grateful to Stefan Landsberger for this point.

informed by Maoism that makes *shanzhai* in China stand out from other forms of pirating around the world. For the latter, it is about today's proletariat inheriting the mass creativity doctrine favoured by Mao and carried out by their predecessors. To say the least, they have made these *shanzhai* phones by adopting practices common in the GLF, such as informal work organisation and indigenous innovation that embodied 'greater, quicker, better and cheaper'.

That said, the downside of inheriting GLF-style mass creativity by contemporary *shanzhai* is also notable, as it has led to numerous failures, similar to that of mass innovation in the wake of the GLF. Previous research on *shanzhai* often focused on its playfulness. However, failure is also part of the *shanzhai* story. Many *shanzhai* phones did not last long as a result of disregarding conventional quality control measures. Some of them even exploded due to cutting corners on materials.

Second, the Maoist governing principle of 'from the masses, to the masses' is also palpable in contemporary *shanzhai*. *Shanzhai* phones with a fancy appearance exemeplify this – one that looks like a model Ferrari sports car, or one made to resemble a pack of luxury cigarettes. Although such designs may be at odds with elite taste, these objects do reflect mass desire. They are popular among the budget-conscious stratum, such as migrant workers who are eager to join the ranks of China's conspicuous consumer society. The novel exteriors of these *shanzhai* phones also look appealing to the middle class and the youth – for whom the possession of one of these phones conveys the desire to be cool and distances them from the dull and generic symbols of consumer society produced by powerful market dominators.

The contemporary proliferation of *shanzhai* is also partly the result of intensive community knowledge sharing, a culture promoted throughout the Mao era. For *shanzhai* phones, protection for intellectual property is weak since design and production revolve around informal work organisation rather than discrete copyright holders. To compete against established manufacturers on turnaround, *shanzhai* makers rely on a model marked by flexibility and informality, thus reducing the importance of formal contracts.[56] Sometimes, in response to rapidly changing market trends, they have even hastily produced phones without a brand name, let alone intellectual property. Despite operating at the border of legality, these informal arrangements have created considerable potential for fostering community learning and knowledge spillover, which will, in turn, upskill the *shanzhai* workforce.

56 Keane, *Creative Industries in China: Art, Design and Media* (Hoboken: John Wiley & Sons, 2013b).

Lastly, the voluntarist mentality of the GLF is still present in contemporary *shanzhai* and it has been renewed with a nationalist tinge. By voluntarist mentality, I mean here the persistent belief among *shanzhai* makers that it is possible to counter a dominant player in marketing and technological sophistication by fully mobilising the creativity, initiative, and entrepreneurial ethos of the Chinese people. Consequently, it is not uncommon for *shanzhai* makers to make such bold assertions as 'world-leading' when marketing their products, despite apparent technological inferiority.[57] Their dauntless attitude towards more powerful rivals is not without grounds, as a few mass created models have turned out to be successful, so much so that even established manufacturers have adopted these mass innovations from their copyists. The dual sim mobile phone, for example, initially a cost-saving design typical of *shanzhai*, has now become an industrial standard implemented by all major phone manufacturers. It exemplifies Abbas's argument that 'the problem of the contemporary fake is not how close the fake is to the original but how close the original is to the fake'.[58] These remarkable cases that have testified *shanzhai's* enduring creative potential are accompanied by nationalist claims, for instance, 'Together let's create the glory of domestic-brand mobile phone industry'.[59] Some even make a connection between present-day *shanzhai* and People's War.[60] People's War, a popular slogan in the Mao era, means to prevail over powerful enemies not by hardware predominance but by overwhelming them in number and with creative tactics. '*Shanzhai* phone makers are able to wage a People's War against powerful competitors such as Nokia, and this is their advantage'.[61] The next section will explore this nationalist dimension of *shanzhai* further.

Self-Reliance: Competing by Copying

In contrast to the previous section, which had a domestic focus and an emphasis on grass-roots creativity, this section takes a different,

57 Youtube, '山寨手機夠晒堅 神奇無線充電 [Magical *Shanzhai* Phone with Wireless Charging Functionality]'.

58 Abbas, 'Faking Globalization', 254.

59 Wallis and Qiu, '*Shanzhai*', 120.

60 IT Time, '"山寨机"出路何在? [What Is the Way out for *Shanzhai* Phones?]', *Sina*, August 5, 2008, accessed August 15, 2015, http://tech.sina.com.cn/t/2008-08-05/13492372897.shtml?qq-pf-to=pcqq.c2c.

61 Ibid.

state-centric perspective and looks at how the state has been actively sponsoring *shanzhai* practices as a means to strengthen China's economic standing internationally. In fact, these two themes are complementary rather than contradictory, as *shanzhai* practices in the Mao era relied upon both bottom-up and top-down initiatives, depending on the circumstances. For example, during such political high tides as the GLF and the CR, mass initiatives were given more prominence; whereas in the politically less turbulent interludes between these pinnacles of mass campaigns, the role of the state was more crucial in leading *shanzhai* efforts in all industries.

To be sure, Mao was only one among many advocates of self-reliance in the history of modern China. Yet, due to international isolation during his time, Mao certainly stood out for having *mobilised* this idea on an unprecedented scale. This was evident in the import substitution industrialisation that he led China to pursue before opening up. This policy of self-sufficiency had not achieved the 'four modernisations' that he and his colleagues hoped for; nonetheless, it made China one of the few countries with a comprehensive industrial structure, without which subsequent reforms would have been impossible. More importantly, his self-reliance doctrine is fertile soil from which to draw inspirational lessons, even in the age of globalisation – such that when China's policymakers realised again the importance of national economic security, the strategies they have implemented are not so different from those of the Mao era. In addition, I argue that *shanzhai* practices in Mao's years have influenced, in a path-dependent manner, China's attempts to upgrade from 'Made in China' to 'Created in China'.

Mao was convinced of the pivotal importance of self-reliance and often attributed the victory of China's communist revolution to the CCP's relative independence from the Comintern and, later, from the Soviet Union. Speaking of the more concrete task of economic construction, the Mao's embryonic thoughts on self-reliance had already emerged during the War of Resistance against Japan, when he established a dozen of Revolutionary Base Areas (again, akin to *shanzhai*) in China's uninhabitable hinterland and turned them into largely self-sufficient economies. The lesson Mao drew from this experience of developing largely without external aid or interference was that:

> We stand for self-reliance. We hope for foreign aid but cannot be dependent on it; we depend on our own efforts, on the creative power of the whole army and the entire people.[62]

This pre-PRC proclamation reveals the essence of Mao's later articulation of self-reliance. For Mao, one's self-independence did not equate with closing oneself off. Rather, it stood for always keeping the initiative in one's own hand, no matter whether foreign assistance was available or not. That is to say, it attached importance to building up the capacity for self-sustainability in the event of adverse situations.[63] This required learning from head to toe the foreign artefact being introduced into China rather than hastily 'transplanting' it without appreciating the pros and cons associated with its local adaptation. Hence, he placed enormous emphasis on the importance of 'sinifying' foreign things. In this respect, he regarded the importation and indigenisation of foreign things as a means to break away from dependence on foreign countries, or to prevent ill-intentioned countries from taking advantage of their technological superiority to extort China's concession. Thus, he was not content with simply putting foreign artefacts into instrumental use or absorbing foreign knowledge alone, but stressed developing and re-innovating further what had been imported.[64] In other words, one could not be said to have fully mastered the skills of making a sophisticated artefact unless (s)he was, on top of reproducing that item, able to modify the original design in order to suit Chinese conditions, and achieve higher efficiency. Mao upheld this view even during the Sino-Soviet honeymoon before the 1960s when production lines along with blueprints were given away to China by the Soviets.

The subsequent Sino-Soviet split and the Soviet withdrawal of assistance certainly reaffirmed the importance of self-reliance. With a heightened degree of isolation as well as victim mentality, China leaned more towards import substitution industrialisation. As a result, the abstract guiding principle of sinification was translated into an emphasis on reverse engineering – an ability that the state invested so much effort into nurturing that it is still a hallmark of Chinese manufacturing today. Reverse engineering as a process of retrospective analysis of structure and design makes quick prototyping possible. It is also a time-wise and cost-saving method compared with

62 Mao, '必须学会做经济工作 [Learning to Handle Economic Work Is a Must]', *News of the CPC*, January 10, 1945, accessed May 15, 2015, http://cpc.people.com.cn/GB/64184/64185/66617/4488951.html.

63 Lieberthal, *Governing China*.

64 Jun Wang, 毛泽东与中国工业化 [*Mao Zedong and China's industrialisation*] (Fujian Education Press, 2001).

traditional mass-production-based design and manufacturing processes.[65] It was therefore not only a rational choice, but a necessity for the isolated PRC to have chosen this shortcut given the flexibility and speediness brought by reverse engineering.

In addition to reversely engineering previously imported artefacts, an accompanying strategy of 'integration innovation' was also deployed. It has also been known as the 'whole-nation system' (举国体制), namely pooling resources of the entire state to tackle problems in a concerted effort. At a time when imports or foreign technical assistance became difficult to obtain, this strategy served as an alternative – a single factory or research institute might not have the technical capability to produce an import substitute on its own, but the obstacle may be partially overcome by assigning each work unit the part of the task at which it excelled and then putting components together to form a complete object. Notably, objects including rockets, satellites, and an atomic bomb, which raised China's international profile, were all made by mobilising resources domestically in the 1960s, during a period when China was most isolated. The Chinese people spoke of these objects using the prefix: *zhengqi* (争气) – credit winning. For example, *zhengqi dan* (争气弹), literally meaning nuclear bomb solely made without external assistance, and which brought credit to China's standing. Hence reverse engineering, integration innovation, and other essentially *shanzhai* practices were justified and glorified as *independent* innovation, because these measures were widely implemented amid international isolation.

Having suffered hardship after breaking away from the Soviets and having learnt from the experience of developing import substitution industrialisation, when external assistance became available again in the early 1970s, the Chinese leadership was even more convinced that importation should only serve as a means to achieve a higher level of self-reliance. Zhou Enlai, for instance, further elaborated Mao's thought on the relationship between foreign import and domestic innovation as a progressive sequence: learn, utilise, modify, and innovate (一学，二用，三改，四创).[66] This laid the foundation for contemporary Chinese thinking on innovation.

For a time after Mao, the emphasis of self-reliance was overshadowed by economic reform as import substitution industrialisation gave way to export-oriented development. The argument was based on the notion of 'comparative advantage' – that China should focus on labour-intensive

65 Vinesh raja and Kiran J. Fernandes, eds., *Reverse Engineering*. Springer Series in Advanced Manufacturing (London: Springer, 2008).

66 Wang, 毛泽东与中国工业化 [*Mao Zedong and China's industrialisation*], 364.

production by taking advantage of low labour costs. In this respect, having a self-sufficient industrial system and being able to produce everything was no longer important, nor desirable. However, major disruption to China's opening up, such as Tiananmen and the subsequent Western embargo, also remind the Chinese leadership that, although international isolation has become a thing of the past, external prejudice against China persists.[67]

A major resurgence of self-reliance thinking occurred in 2008, after the global financial crisis. China realised its exposure to external market turmoil and revaluated export-led development strategy. The outcome has been a paradigm shift from export-oriented growth to innovation-driven development, or 'transformation and upgrade' (转型升级). The 'new' discourse of innovation pronounced by the Chinese government – a. original innovation (原始创新); b. integration innovation (集成创新); c. import, assimilate, absorb, and re-innovation (引进、消化、吸收、再创新) – echoes strongly (especially b and c) with state-led *shanzhai* practices in the Mao era.

In Lieu of a Conclusion: Towards Varieties of *Shanzhai*?

Having discussed *shanzhai* and its various forms in art, manufacturing, and industrialisation, respectively as a mobilisation vehicle for social equality, techniques of mass creativity, and strategy for industrialisation, it must be admitted that instead of making *shanzhai* generalisable, this article has, in fact, further muddied the water and made its definition more opaque. Hence the question, is *shanzhai* really definable? As we have seen, *shanzhai* is like a kaleidoscope – each of its manifestations has provided input to the discourse of *Shanzhai* 'in ways that are specific to its characteristics as a communicative form'.[68] As such, it is perhaps better to speak of 'varieties of *shanzhai*' instead of giving it a rigid definition. This more open-ended approach actually allows more imagination for this evolving phenomenon.

This is not to say, however, that each expression of the *shanzhai* phenomenon is discrete from the other. To the contrary, they do have something in common if viewed through the lens of Maoism. Namely, *shanzhai* as a kind of social innovation creates a participatory space that allows the involvement of previously precluded, non-traditional participants in establishment activities. *Shanzhai* lowers the entry barriers for such establishment activities

67 Jinghan Zeng, Yuefan Xiao, and Shaun Breslin, 'Securing China's Core Interests: The State of the Debate in China', *International Affairs* 91.2 (2015): 245-66.
68 Yang, *Faked in China*, 89.

or even overthrows them altogether. It democratises the right of access for those who 'have not' in art, consumption, and science and technology. Indeed, if one is to characterise *shanzhai* activities in terms of Maoism and its key tenets, they form a coherent set.

Especially on this last point, few other explanatory tools better annotate *shanzhai's* democratising and innovative potential than Mao's Mass Line dictum 'to popularize and improve'(普及和提高) in literature, art, and technology. This is the shared purpose of *shanzhai* regardless of its expressive forms – be they parody, fakery, or copy.

This research is supported by a consolidator grant from the European Research Council (ERC-2013-CoG 616882-ChinaCreative).

Bibliography

Abbas, Ackbar. 'Faking Globalization'. In *Other Cities, Other Worlds: Urban Imaginaries in a Globalizing Age*, ed. Andreas Huyssen (New York: Duke University Press, 2008), 243-66.

Bai, Ye. '山寨文化纵横谈 [Open discussion on *shanzhai* culture]', 社会科学 论坛 (学术评论卷) [*Tribune of Social Sciences*] 2 (2009): 56-90.

Benson, Linda. *China since 1949* (London; New York: Routledge, 2016).

Breslin, Shaun. *Mao: Profiles in Power* (Harlow: Longman, 2000).

Chen, Guisheng. 教育研究空间的探求 [*Explore the space for education and research*] (Fujian Education Press, 2006).

Chubb, Andrew. 'China's *Shanzhai* Culture: "Grabism" and the Politics of Hybridity', *Journal of Contemporary China* 24.92 (2015): 260-79.

Crane, Diana. 'Culture and Globalization: Theoretical Models and Emerging Trends'. In *Global Culture: Media, Arts, Policy, and Globalization*, eds. Diana Crane, Nobuko Kawashima, and Ken'ichi Kawasaki (London: Routledge, 2002), 1-28.

Deng, Qiaobin, and Zhongzhu Jiao. '封建社会的君主专制与文艺功能论 [On autocratic monarchy and the function of literature and art in the feudal society]'. 浙江大学学报(人文社会科学版) [*Journal of Zhejiang University* (*Humanities and Social Sciences*)] 33.1 (2003): 97-104.

Dutton, Michael. 'Maoism and the Structured Intensities of Revolutionary Govern mentality'. In *China's Governmentalities: Governing Change, Changing Government*, ed. Elaine Jeffreys (London: Routledge, 2009), 24-37.

Farmer, Steven M., Pamela Tierney, and Kate Kung-McIntyre. 'Employee Creativity in Taiwan: An Application of Role Identity Theory', *The Academy of Management Journal* 46.5 (2003): 618-30.

Hennessey, William. 'Deconstructing *Shanzhai* – China's Copycat Counterculture: Catch Me If You Can', *Campbell Law Review* 34.3 (2012): 609-60.

IT Times, '"山寨机"出路何在? [What Is the Way out for *Shanzhai* Phones?]', *Sina*, August 5, 2008. Accessed August 15, 2015. http://tech.sina.com.cn/t/2008-08-05/13492372897.shtml?qq-pf-to=pcqq.c2c.

Keane, Michael. *China's New Creative Clusters: Governance, Human Capital and Investment* (London: Routledge, 2013a).

—. *Creative Industries in China: Art, Design and Media* (Hoboken: John Wiley & Sons, 2013b).

de Kloet, Jeroen, and Lena Scheen. 'Pudong: The *Shanzhai* Global City', *European Journal of Cultural Studies* 16.6 (2013): 692-709.

Lieberthal, Kenneth. *Governing China: From Revolution Through Reform*. 2nd edition. (New York: W. W. Norton & Company, 2004).

Mao, Zedong. 'The Mass Line'. In *Quotations from Mao Tse Tung* (Peking Foreign Languages Press, 1966). Accessed May 11, 2015. https://www.marxists.org/reference/archive/mao/works/red-book/ch11.htm.

—. '论十大关系 [On the Ten Major Relationships]', *Xinhuanet*, December 30, 2004, 1956. Accessed June 10, 2015. http://news.xinhuanet.com/ziliao/2004-12/30/content_2393996.htm.

—. '必须学会做经济工作 [Learning to Handle Economic Work Is a Must]', *News of the CPC*, January 10, 1945. Accessed May 15, 2015. http://cpc.people.com.cn/GB/64184/64185/66617/4488951.html.

—. '在延安文艺座谈会上的讲话 [Talks at the Yan'an Forum on Literature and Art]', *News of the CPC*, May 1942. Accessed May 15, 2015. http://cpc.people.com.cn/GB/64184/64185/66617/4488965.html.

Meisner, Maurice. *Mao Zedong: A Political and Intellectual Portrait* (Cambridge: Polity, 2007).

Pang, Lai-kwan. *The Art of Cloning: Creative Production during China's Cultural Revolution* (London: Verso, 2017).

People's Daily. '上海的技术革命群众运动 曹荻秋代表的发言 [The Mass Campaign of Technological Revolution in Shanghai – Speech by Representative Cao Diqiu]', April 9, 1960.

—. '社论: 把总路线的红旗插遍全国 [Editorial: Stick the Red Flag of the General Line All over the Country]', May 29, 1958a.

—. '土法炼钢 妙计无穷 [An Unlimited Number of Ingenious Techniques That Were Indigenously Developed to Make Steel]', October 5, 1958b.

Raja, Vinesh, and Kiran J. Fernandes, eds. *Reverse Engineering*. Springer Series in Advanced Manufacturing (London: Springer, 2008).

Rudowicz, Elisabeth, and Xiao-dong Yue. 'Compatibility of Chinese and Creative Personalities', *Creativity Research Journal* 14.3-4 (2002): 387-94.

Shao, Zenghua. 韓非子今註今譯 [*Contemporary annotation and interpretation of Han Fei Zi*] (Taipei: Commercial Press Taiwan, 1990).
Sigley, Gary. 'Chinese Governmentalities: Government, Governance and the Socialist Market Economy', *Economy and Society* 35.4 (2006): 487-508.
Wallis, Cara, and Lin-chuan Qiu. '*Shanzhaiji* and the Transformation of the Local Mediascape in Shenzhen'. In *Mapping Media in China: Region, Province, Locality*, eds. Wanning Sun and Jenny Chio (London and New York: Routledge, 2012), 109-25.
Wang, Jun. 毛泽东与中国工业化 [*Mao Zedong and China's industrialisation*] (Fujian Education Press, 2001).
Wong, Winnie Won-yin. *Van Gogh on Demand: China and the Readymade* (Chicago, IL; London: University of Chicago Press, 2014).
Xiang, Dongmin, and Yihui An. '吹响"大跃进"冲锋号的中共八大二次会议 [The CCP's 8th Congress 2nd Session that launched the Great Leap Forward]', 文史精华 [*Gems of Culture and History*] 3 (2011): 24-32.
Yang, Fan. *Faked in China* (Bloomington, IN: Indiana University Press, 2015).
—. 'China's "Fake" Apple Store: Branded Space, Intellectual Property and the Global Culture Industry', *Theory, Culture & Society* 31.4 (2014): 71-96.
YouTube. '山寨手機夠晒堅 神奇無線充電 [Magical *Shanzhai* Phone with Wireless Charging Functionality]', 2013. Accessed August 15, 2016. https://www.youtube.com/watch?v=eX--pvEWNso.
Yu, Hua. 十个词汇里的中国 [*China in Ten Words*] (Taipei: Rye Field Publishing Co., 2011).
Zeng, Jinghan, Yuefan Xiao, and Shaun Breslin. 'Securing China's Core Interests: The State of the Debate in China', *International Affairs* 91.2 (2015): 245-66.
Zhang, Lin, and Anthony Fung. 'The Myth of "*Shanzhai*" Culture and the Paradox of Digital Democracy in China', *Inter-Asia Cultural Studies* 14.3 (2013): 401-16.
Zhu, Yunhe. '从"土洋并举"到"以土为主" – "大跃进"时期技术革命方法的嬗变 [From "combining indigenous and foreign techniques" to "mainly rely on indigenous techniques" – the evolution of methods deployed in technological revolution during the Great Leap Forward]', 北京党史 [*Beijing Party History*], 6 (2011): 26-31.

About the Author

Xiao Yuefan was Postdoctoral Research Fellow at the Department of Media Studies, University of Amsterdam. He has a PhD in Politics and International Studies from the University of Warwick, focusing on the politics of crisis management in China. He previously worked for the ERC-funded project ChinaCreative, for which he analysed how governmentality was entangled with the development of creative industries in China.

2.4 'Isn't that funny?'

The Unsettling Effect of *Shanzhai* Products

Lena Scheen

Abstract
This essay questions the fascination with *shanzhai* products, in particular by people from advanced economies. Why do *shanzhai* products make us laugh? What does this tell us about our shared assumptions on brands and their symbolic meanings? Or about the newness dogma dominating Euro-American cultures? Instead of focusing on the creative potential of *shanzhai*, this essay highlights the liberating effect of blunt copying.

Keywords: *shanzhai*, self-symbolism, cultural superiority, copying, creativity

It was a five-storey building in Shanghai's Xujiahui district. On each floor, a noisy crowd of people strolled through a labyrinth of makeshift shops selling electronic products. I wanted to buy an iPhone and had asked a Shanghai friend to help me with the bargaining. We pushed through the crowd to a little shop whose walls were covered with Apple logos. Getting closer, I noticed something odd: the logos missed the little bite in the apple that prominently marks each Apple product. Indeed, the brand name was also different: in the same font Apple uses, it read Pingguo instead, the *pinyin* transcription of the Chinese characters for 'apple'.

Laughing, I turned to my friend: 'look, they're fake iPhones! Isn't that funny?' Not understanding my amusement, my friend drily corrected me, 'No, it's not Apple, it's Pingguo, their mobiles are much cheaper than iPhones and they have many more functions'. The seller quickly chimed in explaining in great detail which features were the same as the iPhone's, and which were even better. Like most foreigners in China, I responded to this blunt comparison between the 'real' and the 'fake' with a mixture of amusement, awkwardness, and fascination.

Like the brands HiPhone, iPear, X-Apple, and iOrgane (an intentional misspelling of iOrange), the Pingguo is a prime example of a *shanzhai* iPhone: it looks like an iPhone and has a brand name and logo that clearly refer to the iPhone, but it is not an exact copy of an iPhone meant to deceive the customer into thinking s/he is buying a 'real' iPhone. In other words, it is not 'fake' – *jiade* in Chinese – which is why my friend protested against my use of the words 'fake iPhone', while he didn't see a problem in buying a *shanzhai*. A *shanzhai* product is meant to be similar enough to its 'prototype' to give the customer positive associations with the original brand, but different enough to compete with it and to avoid – with varying success – violating intellectual property rights (IPR).

When my American or European colleagues and I are in one of Shanghai's many shopping centres, we point at *shanzhai* shops and products and, without having to say anything, we look at each other and laugh. While seemingly innocent, these moments of mutual understanding reveal shared assumptions that treat a common practice in China as something odd or funny; something worth pointing at and laughing about. *Shanzhai* products are also the subject of countless blog posts and magazine articles 'recounting the silliness bordering on infantilism of the not-yet-assimilated Chinese', as Barton Beebe observes.[1] Our laughter thus betrays a sense of cultural superiority and, at the same time, as Beebe also argues, 'the same innumerable blog posts that seem to make fun of *shanzhai* also typically betray a degree of admiration for the audacity and eccentricity of the conduct'.[2] Today, there is even an increasing scope of scholarship on *shanzhai* products exploring not only its legal and economic implications, but also its creative and innovative potential. All this raises the question why *shanzhai* products have this unsettling effect on us – in particular people from advanced economies – whether it being amusement, awkwardness, or fascination.

Many scholars have shown how, in 'high modernity',[3] consumers buy products no longer solely for their utility value but for their symbolic

1 Barton Beebe, 'Shanzhai, Sumptuary Law, and Intellectual Property Law in Contemporary China', *47 U.C. Davis Law Review* (2014): 870.

2 Ibid., 872.

3 Anthony Giddens, *Modernity and Self-identity: Self and Society in the Late Modern Age.* (Cambridge: Polity Press, 1991).

Kloet, Jeroen de, Chow Yiu Fai, and Lena Scheen (eds), *Boredom,* Shanzhai, *and Digitisation in the Time of Creative China.* Taylor & Francis Group, 2019
DOI: 10.5117/9789462984745_CH2.4

meanings in the construction, maintenance, and communication of self and group identity. The functions of the symbolic meanings of products operate in two directions, as Richard Elliott has argued, 'outward in constructing the social world, *social-symbolism*; and inward towards constructing our self-identity, *self-symbolism*'.[4] Brands play a particularly important role in this process of self-identification as, in the words of Elliott and Wattanasuwan, 'the symbolic consumption of brands can help establish and communicate some of the fundamental cultural categories such as social status, gender, age, and such cultural values as family, tradition and authenticity'.[5] In their study on brands as symbolic resources for the construction of identity, Elliott and Wattanasuwan, also show how brands can acquire deep meaning for consumers by their involvement in the socialisation process of growing up.[6] In particular, the brands people consumed in their childhood evoke strong feelings of nostalgia and provide a sense of comfort in times of insecurity.

These symbolic meanings of brands, on a social, psychological, and emotional level, help explain people's uneasiness with products that seem to ridicule the brand names they consume and strongly identify with. One could say that *shanzhai* are intruding on holy ground. As shown by the *shanzhai* iPhone examples above, one of the trademarks of the *shanzhai* industry is their clever choices of 'pun brand names'; other examples are Bew Nalance for New Balance, Star Fucks and Sunbucks for Starbucks, Kuma and Punk for Puma, or Hike, Like, and Dike for Nike. In addition, the logos of *shanzhai* products are characterised by playful references to the original brands; such as a pear (both with and without a bite) or an orange on *shanzhai* iPhones, or a puma animal with a mohawk haircut on Punk clothing. Counterintuitively, the fact that *shanzhai* products are cheaper does not necessarily mean that they are of lower quality. On the contrary, as business magnate Jack Ma (founder of China's largest e-commerce company Alibaba Group) asserts: 'The problem is that the fake products today, they make better quality, better prices than the real names. The exact factories, the exact same materials, but they do not use their names'. In other words, *shanzhai* products are not just challenging economic and legal regulations, they openly make fun of the original famous brands: they play with their names and logos, randomly add or remove features, are always cheaper and sometimes even better than the brands they copy.

4 Richard Elliot, 'Symbolic Meaning and Postmodern Consumer Culture'. In *Rethinking Marketing: Towards Critical Marketing Accountings*, ed. Douglas Brown (Thousand Oaks, CA: Sage, 1999), 113.

5 Elliot and Kristsadarat Wattanasuwan, 'Brands as Symbolic Resources for the Construction of Identity', *International Journal of Advertising*, 17.2 (1998): 139.

6 Ibid., 140.

My personal favourite *shanzhai* product is the BlockBerry, a look-a-like BlackBerry that masters all the *shanzhai* characteristics. The mobile's advertisement features President Obama in a suit with an American-flag pin (ironically or accidentally?) backwards on the wrong lapel.

Made in 2009, the makers evidently responded to the news of Obama being the first president of the USA to demand a BlackBerry, revealing that they anticipated their targeted Chinese customers picking up on this reference as well. Since President Obama famously owned a BlackBerry Storm 9500, the makers came up with the fitting name BlockBerry Whirlwind 9500 and the advertisement slogan 'Obama's BlackBerry. My BlockBerry Whirlwind 9500'. According to the advertisement, the mobile is created by the company HAFF-COMM, *hafo tongxin* (哈佛通信). The literal translation of the Chinese name *hafu tongxin* is Harvard Communication, a clear reference to the university where president Obama graduated from Law School. In the words of a Chinese netizen commenting on pictures of *shanzhai* products: 'Who says the Chinese lack creativity?'

In this volume, Eitan Wilf shows how creative practices in China problematise the Western-Modern-Romantic notion of creativity in which it is always *contrasted* with imitation, routinisation, and reproduction. Whereas, as Wilf argues, imitation is in fact 'part and parcel' of creative practice, as the acquisition of skills, first, requires imitation. Second, to creatively break away from conventions one has to first inhabit them via imitation. Third, imitation is inherently creative to some degree.[7] In the Western-Modern-Romantic imagination an 'authentic' artwork is created by an artist who is a 'highly exceptional individual'.[8] Both the artist and the work are supposed to be 'unique'. Few would regard a painter copying a Van Gogh painting in the '*shanzhai* Mecca' of Dafen Oil Painting Village 'an artist' or his painting an 'artwork', even if he made substantial changes to the original.

Whether we talk about an artwork, a mobile phone, or an academic article, what Euro-American society – legally and culturally – demands is that we do not imitate, reproduce, or copy, but that we create something *new*. An artist should make a *new* artwork, a scholar should do a *new* discovery, make a *new* argument, develop a *new* theory. But, as most scholars would admit, our scholarship is little more than a collection of other people's ideas, put in a new order, a new framework, and with a new title added. Like Wilf argues on the creative practice of art, imitation is also 'part and parcel' of

7 See Wilf in this volume.

8 Ibid.

2.4.1 The mobile's advertisement features President Obama in a suit with an American-flag pin (ironically or accidentally?) backwards on the wrong lapel

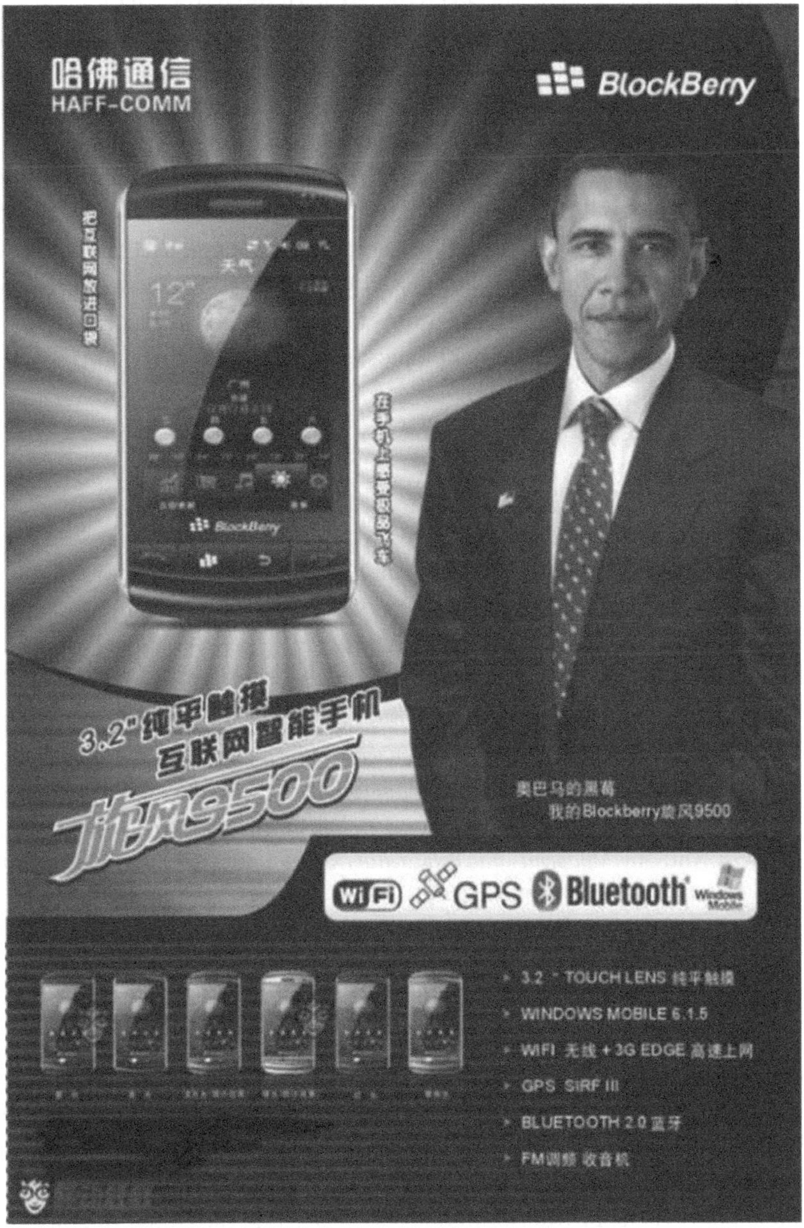

scholarly work. While we acknowledge the 'originals' in our references, we claim their ideas have inspired us to new ideas that are vitally our own.

The only *real* difference between the 'Western-Modern-Romantic' notion of an artwork, a mobile, or a scholarly article, and a *shanzhai* product is in its *intention*, in its boldness, in its unpretentious copying of the original. A *shanzhai* does not try to conceal the original it copies, instead it takes a brand product and changes it just enough to make a new product. While many have pointed at the creative and innovative process of these careful *changes*, I want to highlight the liberating effect of its blunt *copying*. What makes *shanzhai* truly 'unique' is precisely that it is not unique; that it refuses to pretend its uniqueness, its authenticity, its newness. A *shanzhai* resists the newness dogma dominating Euro-American cultures. Instead, it screams in our faces: 'yes, I'm a copy, but I'm better and I'm proud of it.'

Bibliography

Beebe, Barton. 'Shanzhai, Sumptuary Law, and Intellectual Property Law in Contemporary China', *47 U.C. Davis Law Review* (2014): 849-74.

Elliott, Richard. 'Symbolic Meaning and Postmodern Consumer Culture'. In *Rethinking Marketing: Towards Critical Marketing Accountings*, ed. Douglas Brown (Thousand Oaks, CA: Sage, 1999), 112-25.

Elliott, Richard, and Kritsadarat Wattanasuwan. 'Brands as Symbolic Resources for the Construction of Identity', *International Journal of Advertising*, 17.2 (1998): 131-44.

Gabriel, Yannis, and Tim Lang. *The Unmanageable Consumer: Contemporary Consumption and its Fragmentations* (London: SAGE Publications, 1995).

Giddens, Anthony. *Modernity and Self-identity: Self and Society in the Late Modern Age*. (Cambridge: Polity Press, 1991).

About the Author

Lena Scheen is Assistant Professor of Global China Studies at NYU Shanghai, and Global Network Assistant Professor at NYU. Scheen's research explores the social and cultural impact of China's fast urbanisation, focusing on Shanghai and storytelling. Her publications include the monograph *Shanghai Literary Imaginings: A City in Transformation* (AUP, 2015) and the edited volume, together with Jeroen de Kloet, *Spectacle and the City: Chinese Urbanities in Popular Art and Culture* (AUP, 2013). She is currently working on her new manuscript *Shanghai Storytelling: City Myths, Historical Memory, and Place Attachment*.

2.5 *Shanzhai* = Creativity, Creativity = *Shanzhai*

Stefan Landsberger

Abstract

Shanzhai has come to refer to all Chinese counterfeit products and to the attitude of their producers towards authority. One argument that explains the popularity of *shanzhai* products in China is that the average consumer could not afford original name brands, and that by acquiring a cheaper version, one could find out whether it was worthwhile to own an original. While *shanzhai* has turned into a phrase and concept that many in the West are familiar with, many Chinese are not, or no longer; most do not care whether the goods they consume are 'authentic' or not. Other Chinese are embarrassed by the frequent use of the concept in combination with the signifier 'China'.

Keywords: *Shanzhai*-as-mimicry, counterfeit, authenticity, Dafen Oil Painting Village, Dark *shanzhai*

Literally, *Shanzhai* refers to the mountain stronghold defended by fiercely patriotic freedom fighters à la Robin Hood, resisting the authoritarian imperial rule of the Song Dynasty and supporting the poor and the weak. The image has been around for a long time in China. Shi Nai'ans popular novel *Water Margin*, which was first published in 1589, had made the rebels, as well as their stronghold in the Liangshan Marshes, household names in China. Non-Chinese readers had to wait until the 1930s to get to know these guerrillas and learn of their exploits, when Pearl Buck's masterful translation *All Men are Brothers* (1933) appeared.[9] Over time, *shanzhai* came to refer to all counterfeit products and to the attitude of their producers towards authority. After China's accession to the World Trade Organization in 2001, the flood of goods coming out of China that had already started to

9 Fan Yang, 'From Bandit Cell Phones to Branding the Nation: Three Moments of *Shanzhai* in WTO-era China', *Positions: East sia cultures critique* 24.3 (2016): 590; William Hennessey, 'Deconstructing *Shanzhai* – China's Copycat Counterculture: Catch Me If You Can', *Campbell Law Review* 34.609 (2012): 614-616.

make inroads in the 1980s, increased exponentially. By 2008, the 'year of *shanzhai*' as the internet site *Renmin wang* (People's Net) named it,[1] when China monopolised the global gaze through the Olympics, the Wenchuan Earthquake, the riots in Tibet and Xinjiang, the Sanlu Milk Powder Scandal, and its first space mission, *shanzhai* had become the moniker, both inside China and without, for cheap imitation products, copycat wares. China came to be seen as the bootlegging capital of the world.[2]

Is *Shanzhai* 'Typically' Chinese?

Much has been theorised about the origins of *shanzhai*-as-mimicry, as counterfeit, often pointing to the differences in appreciating creativity (and originality) that are said to exist between China and the West (or the rest). The Chinese traditionally valued technical expertise as expressed in immaculate workmanship, attained by endlessly copying a flawless original created by an acknowledged master. In other words, the ability to masterfully copy was seen as proof of a creator's artistic value. The quality of the final product mattered more than whether it was an original piece of work or not. As Fan Feng argues in this volume, the West, for example the Delft Blue pottery producers in Holland, *shanzhaied* original Chinese craftsmanship. This indicates that copying is not necessarily an activity that can be monopolised by a single people or nation but is part of a wider developmental process.[3] On a more contemporary note, Xiao Yuefan demonstrates in this volume how the roots of the practice of *shanzhai* actually could be traced back to important discursive elements in Mao Zedong Thought (i.e. the concepts of art for the people and the Mass Line and the principle of self-reliance) that were translated into practice in such mass movements as the Great Leap Forward. On the basis of such arguments, copying has been defended as an inherent trait of Chinese culture, and *shanzhai* production as something to be fiercely proud of.

1 Yang, 'From Bandit Cell Phones to Branding the Nation', 589.

2 Lai-kwan Pang, 'China Who Makes and Fakes – A Semiotics of the Counterfeit', *Theory, Culture & Society* 25.6 (2008): 120.

3 Ibid., 123,

Kloet, Jeroen de, Chow Yiu Fai, and Lena Scheen (eds), *Boredom,* Shanzhai, *and Digitisation in the Time of Creative China*. Taylor & Francis Group, 2019
DOI: 10.5117/9789462984745_CH2.5

But *shanzhai* was – and is – more than just industrial-scale counterfeiting and piracy; it consists of 'parodies, irreverent protests, and "grassroots innovations" that exploit "the ambiguities" and skirt the rules rather than break them'.[4] This precisely puts a fine point on the different ways of interpreting *shanzhai*: many non-Chinese see *shanzhai* products merely (and only) as 'simply low-priced counterfeits of well-known brands',[5] completely missing the playfulness that is part and parcel of the process. Equating *shanzhai* with counterfeit also turns a blind eye to the elements of adopting global goods, grass-roots participation, and resistance against authority that are bound in the original meaning of the term.

Shanzhai as Emblem of Development and Modernity

The term *shanzhai* as it emerged in the 2000s initially was very much associated with cheap mobile phones, often produced by small, semi-legal workshops, many of them family-run, which imitated established (foreign) brands. These producers made use of the opportunities provided by access to (more) advanced technology from elsewhere, the opportunity to run private businesses and sheer entrepreneurship. Their phones first emerged in Southeast China's Pearl River Delta, comprising Guangzhou, Hong Kong, and the adjacent bubble-city of Shenzhen, the region that became the crucible of the reform and development frenzy that has propelled China forward. The qualities these phones offered were multiple; they were cheap, for sure, and this allowed many, often with little money in their pockets, to join the wonderful world of mobile communications. It is no wonder that the largest initial group of users could be found among the migrant workers who had moved from the countryside to the cities, providing the cheap fuel and lubricants, as it were, for the Chinese economic machine.[6] At the same time, they conferred status on their users, not only because they looked almost exactly like the much more expensive originals that they copied,[7] but because they allowed access to advanced modes of communications. Moreover, they often could boast of functions that the originals did not

4 Hennessey, 'Deconstructing *Shanzhai*', 660-61.

5 Yiying Wu and Jack Whalen, *Portraits of Chinese Shanzhai* (New York: American Sociological Association Annual Meeting 2013).

6 Carolyn Cartier, Manuel Castells, and Lin-chuan Qiu, 'The Information Have-Less: Inequality, Mobility, and Translocal Networks in Chinese Cities', *Studies in Comparative International Development* 40.2 (2005): 11; Yang, 'China's "Fake" Apple Store: Branded Space, Intellectual Property and the Global Culture Industry', *Theory, Culture & Society* 31.4 (2014): 83-84.

7 Wu and Whalen, *Portraits of Chinese Shanzhai*.

have (yet) and increasingly, they came with a quirk. Just think of the mobile phone unit with a Rolex Oyster watch plate attached to it, combining two symbols of taste, refinement and wealth into one invincible statement of status. I personally find the packet-of-Panda-cigarettes-phone one of the greatest inventions ever; all the more so because Panda cigarettes were the favourite brand of Deng Xiaoping, the 'Chief Architect' of China's reform and opening up. This particular phone thus not only let its users have the vicarious pleasure of owning a pack of Panda cigarettes, but at the same time supported and/or paid homage to the leader who had made owning a mobile phone possible in the first place.

As more companies came in contact with foreign products and companies for which they started to assemble semi-finished products in the 1990s, the menu of *shanzhai* products grew. As Chubb writes, '[f]rom *shanzhai* mobile phones to *shanzhai* superstars, from *shanzhai* news to *shanzhai* police stations, from *shanzhai* cigarettes to *shanzhai* trains, for a time it seemed that nothing could not be *shanzhai*ed in China'.[8] It needs further research to ascertain whether the Adidos or Dasabi brand running shoes on offer were thus named on purpose, to evade copyright issues; to hoodwink customers who are unaware of the difference; or with an ironic wink to the more knowledgeable consumer. In the same vein, does the customer wearing clothes from the Dolce & Banana brand know they are the real deal or not? Does a cup of Starfucks coffee give the same satisfaction as its (almost) namesake? And does it really matter? One argument explaining the popularity of *shanzhai* products frequently heard in China in the 1990s and early 2000s was that original brands were out of reach for the average Chinese consumer, and that by acquiring a cheaper version of them, s/he would be able to find out whether it was worthwhile to aspire to owning an original in the future or not.

The Legality of *Shanzhai*

Increasingly, however, copyright holders started to make noises about missing out on their royalties. As long as these companies were located outside of China, and had deep pockets, the Chinese authorities did not really seem to care. Yes, of course, it could be heard in official quarters, China was a nation one can depend on; of course, China observed its legal obligations; of course, copyright was something sacred in China too; certainly, China

8 Andrew Chubb, 'China's *Shanzhai* Culture: "Grabism" and the politics of hybridity', *Journal of Contemporary China* 24.92 (2015): 261.

would do its utmost to apprehend and punish the perpetrators. The problem was that '[S]*hanzhai* behaviour is not necessarily against the law; it is just outside of the government's control'.[9] This applied all the more when the *shanzhaiing* companies in question operated under the actual and active protection of state agencies. It is worth noting that as soon as Chinese companies discovered that their own products came to be '*shanzhaied*' by local companies in more remote inland areas, a considerable amount of official effort could be freed to try and bring these practices under control, thereby also benefitting the foreign companies that had been complaining about the practice all along. In the process, the finer ins-and-outs of copyrights and their protection became better understood.

While this process took form, many of the small companies that originally had started out as *shanzhai* producers attained respectability, evolving into legitimate businesses that took on established global brands on the basis of original new products, often in completely different fields from where they had started. The BYD Company (Build Your Dreams) is a good example of this process. It started out producing rechargeable batteries for mobile phones, competing with and potentially making use of (i.e. *shanzhaiing* technology developed by) Japanese companies, which monopolised the Chinese market. After 2002, it branched out into automobile research and development, becoming one of the big electric and gas-powered auto makers in China. In a way, something similar has happened in respect to the Dafen Oil Painting Village. First, it was merely a place where painters worked, copying established masterpieces, initially for the European and American markets, later also for the domestic one. It attracted curious foreign journalists, who wrote humorous media reports about fake Van Gogh's. Then, it became an example of a village-based cultural industry that merited emulation elsewhere in China, as can be seen from the fact that it was presented as the main theme of the Shenzhen Pavilion in the Best Urban Practices Zone during the Shanghai Expo in 2010.[10] Now it has become a *Gesamtwerk*: it is like a huge studio where art is recreated in various small workshops by professionals, and at the same time a tourist village providing the spectacle of an art-copying village in full operation. The Dafen Art Museum attributes legitimacy to the art works produced within the city boundaries by exhibiting them for what they are: works of art. Questions about their authenticity or originality are neither raised, nor answered.

9 Hennessey, 'Deconstructing *Shanzhai*', 661.

10 Si-ming Li, Hung-ha Cheng, and Jun Wang, 'Making a Cultural Cluster in China: A study of Dafen Oil Painting Village, Shenzhen', *Habitat International* 41 (2014): 156, 158-160.

Dark *Shanzhai*

Let us not forget that there also is a dark side to what most consider the playfulness of *shanzhai*. I am thinking specifically of the counterfeit goods that are not produced with the intention of selling them cheaply, but to fool the end users into believing they are buying original products and thereby make a fortune. Those falling victim of these practices often tend to be people from the (remote) countryside who lack the savviness of urbanites.[11] Both inside China and outside, cases have been reported about the sales of fake (but expensive) medical drugs; about baby formula made of nothing but chalk; about spare parts used in the aviation and other industries that have not properly gone through quality control regimes, etc. With the lives of innocent people at stake, these cheap production processes only seem to benefit producers without qualms or conscience.

Yet, *shanzhai* as such seems to have become a non-brand, a 'not-good' product because it has no recognisable brand name.[12] It is interesting to note that while *shanzhai* has turned into a phrase and concept that many in the West are familiar with, many Chinese are not or no longer; or they do not care about the goods they consume being 'authentic' or not. Other Chinese are embarrassed by the frequent use of the word/concept in direct combination with the word 'China'. At the same time, however, as the mood of the nation has taken on a more patriotic hue from the 1990s onwards, more and more people consider buying *shanzhai*, i.e. locally produced goods rather than (expensive) imported ones, as a way of showing that they not only 'love the nation' but actively support its industries, thus expressing a form of consumer nationalism while at the same time being able to enjoy and take part in the pleasures of global consuming.[13] As times have passed, the clear distinctions between real and not, between foreign-produced and indigenous increasingly have become blurred. For many of the post-90s generation and beyond, the so-called millennials, the question of buying (into) *shanzhai* is no longer relevant. They have developed their own playfulness, their own way of mixing-and-mashing-up, and their own behavioural and sartorial codes.

11 Wu and Whalen, *Portraits of Chinese Shanzhai*.

12 Ibid.

13 Lu-wang Cheng and Zhen Xiong Chen, 'Consumer Ethnocentrism and Willingness to Buy Domestic Products in a Developing Country Setting: Testing Moderating Effects', *Journal of Consumer Marketing* 21.6 (2004): 393-394.

In Closing

Given that *shanzhai* is not branded, why is it that until the present, no clever businessperson or marketer has registered a copyright for the term (or the practice of) *shanzhai*? It surely is a great brand name with global recognition. As a brand, it offers wonderful opportunities to produce goods (or services) of whatever stripe, although not necessarily with the same tongue-in-cheek qualities that some Chinese *shanzhai* products (have) had. I find it odd that the Chinese entrepreneurial spirit, the ethos of which is personified in Alibaba founder Jack Ma Yun, who is praised and worshipped on an almost daily basis both within and without China, has had a blind spot for this. So, let's use *shanzhai* as a rallying cry to turn 'China the Pirate; Chinese, the Uncreative' into their opposites! To make it the brand that signifies the turn from 'made in China' to 'created by China'!

Bibliography

Cartier, Carolyn, Manuel Castells, and Lin-chuan Qiu. 'The Information Have-Less: Inequality, Mobility, and Translocal Networks in Chinese Cities', *Studies in Comparative International Development* 40.2 (2005): 9-34.

Cheng, Lu-wang, and Zhen Xiong Chen. 'Consumer Ethnocentrism and Willingness to Buy Domestic Products in a Developing Country Setting: Testing Moderating Effects', *Journal of Consumer Marketing* 21.6 (2004): 391-400.

Chubb, Andrew. 'China's *Shanzhai* Culture: "Grabism" and the Politics of Hybridity', *Journal of Contemporary China* 24.92 (2015): 260-279.

Hennessey, William. 'Deconstructing *Shanzhai* – China's Copycat Counterculture: Catch Me If You Can', *Campbell Law Review* 34.609 (2012): 609-661.

Li, Si-ming, Hung-ha Cheng, and Jun Wang. 'Making a Cultural Cluster in China: A Study of Dafen Oil Painting Village, Shenzhen', *Habitat International* 41 (2014): 156-164.

Pang, Lai-kwan. 'China Who Makes and Fakes – A Semiotics of the Counterfeit', *Theory, Culture & Society* 25.6 (2008): 117-140.

Wu, Yiying and Jack Whalen. *Portraits of Chinese Shanzhai* (New York: American Sociological Association Annual Meeting 2013).

Yang, Fan. 'From Bandit Cell Phones to Branding the Nation: Three Moments of *Shanzhai* in WTO-era China', *Positions: East Asia Cultures Critique* 24.3 (2016): 589-619.

—. 'China's "Fake" Apple Store: Branded Space, Intellectual Property and the Global Culture Industry', *Theory, Culture & Society* 31.4 (2014): 71-96.

About the Author

Stefan Landsberger is Associate Professor at the Department of China Studies and has been part of Leiden University for over thirty years, holding various positions. From 2005-2015, Landsberger held the Olfert Dapper Chair of Contemporary Chinese Culture at the University of Amsterdam. He specialises in contemporary China, from a social and political perspective, and is an expert on Chinese political communication and visual propaganda.

NOTES FROM THE FIELD

2.6 Bringing the Chinese Dream to the U.S.

A Curatorial Practice in Art Education

Feng Fan

Abstract

In 2016, Feng directed an intense academic exchange programme between the Department of Art at the University at Buffalo, and the Academy of Arts & Design at Tsinghua University, Beijing, on both sides of the Pacific. The core of this project was to bring the students from two cities and different cultural backgrounds to each other's cities to make art and foster creativity. Using *shanzhai* as a curatorial strategy, they were able to align the American dream to its Chinese rendition as promoted by Xi Jinping, reviewing the regional history, geographical politics and contemporary social dynamics, embedding them in a broader cultural/ economic context to advance critical speculation: the political, economical and globalised framework of the 'Chinese Dream' ideology.

Keywords: Chinese Dream, *shanzhai*, art pedagogy, contemporary art, intertextuality

'My Chinese Dream is to migrate to the U.S. and realise my American Dream'
Quotation of a high rated answer from Zhihu.com[14] *responding to the question of 'The Chinese Dream and the American Dream, which one is easier to realise'?*

14 A popular Chinese question-and-answer website with high-profile Chinese entrepreneurs and public intellectuals among its users.

Kloet, Jeroen de, Chow Yiu Fai, and Lena Scheen (eds), *Boredom,* Shanzhai, *and Digitisation in the Time of Creative China.* Taylor & Francis Group, 2019
DOI: 10.5117/9789462984745_CH2.6

The idea behind the phrase 'Chinese Dream', which was put forward by the CPC General Secretary Xi Jinping in 2012, is very similar to the idea of the 'American Dream': It stresses entrepreneurship and glorifies a generation of self-made men and women in post-reform China, achieving magnificent improvements in terms of their living standards and social life. The first response of most Chinese who first heard this phrase was that it was a *shanzhai* version of the American dream, stirring a heated debate among the public about the differences between the two concepts, despite the explanations given by the officials. The debate turned into a comprehensive and multi-layered discussion on a series of issues, from the political ideology to the economic system, collectivism to individualism, or more specific topics such as originality and *shanzhai*, personal development and pursuit of lifestyle. Such interpretative structural complexity provides a rich ambiguity full of angles for artistic intervention.

In 2016, I directed an intense academic exchange programme between the Department of Art at the University at Buffalo, and the Academy of Arts & Design at Tsinghua University, Beijing; it took place on both sides of the Pacific. Under the title of 'Beijing/Buffalo: Rust Never Sleeps', the programme was aiming to support biennial cross-cultural collaborations between MFA and Ph.D. students from both universities in the form of artworks, exhibitions, and publications. The core of this project was to bring the students from two cities and different cultural backgrounds to each other's cities to make art and foster creativity.

In a programme that involves the collaboration of artists from the U.S. and China with the final exhibition to be held on each end, the relationship between the exhibiting venues, art projects, and the audience is not naturally given, but must be carefully constructed by a well articulated curatorial plan. If a catalogue or symposium creates an interpretative context for an exhibition, then an exhibition's spatial structure can become a vehicle in developing critical concepts and methodological discourses. Therefore, when curating an exhibition, space is one of the key ways to engage art criticism and theoretical thinking. The concept of 'exhibition space' refers not only to the architectural space but to a larger idea that denotes a dynamic spatial and perceptual network constituted by the physical space, the art project, and the audience: not only does architecture contain an exhibition physically but an exhibition also necessarily redefines the site, endowing a given site with new, specific meaning. Therefore, when the artists from Tsinghua University visited Buffalo, the history of the city and its relationship with Beijing immediately caught my attention.

In the case of this project, the exhibition spaces are always connected to a particular methodological discourse, though by creating these spaces and discourses the educational project automatically engages in a negotiation with art systems and even challenges and reinvents the existing institutional framework of art education and art production.

For example, one installation piece, created by Teng Teng, Zhou Quan, and Liz Black and titled 'I Am in the Dark Here', of a glowing Chinese literati 'mountain' made of traditional Chinese 'Xuan' paper was soaked in a tank filled with a mixture of Chinese Ink and polluted water taken from the Lake Erie. With time, the mountain, a symbol of traditional spiritual value, would collapse by the gradual corroding of the base. What gave the piece hypnotic and reassuring visual pleasure was the video projection on the mountain: a video clip of the hovering foam of the polluted water with the reflection of the sunlight. The almost hallucinatory visual effect of the light and texture made one forget the danger and poisonous nature of the image. One performance piece by Lau Cherry Yung Yung and Morgan Layne, titled 'Double Rule', also involved the motif of water, but by constantly pouring water from two ladders into a pool containing the lake water with mud and debris found at the shore. It highlighted the human interference with nature when the calm and clear water became muddy and dingy. Yet, after the performance was completed, the water in the pool regained its calm and clarity, again with time. The piece showed nature's ability of self-repairing. Another piece done by the art history student Zheng Xinyu and his counterpart Caroline Doherty, titled 'Rule Rules Ruled Ruler', was a political allegory of the Sino-American relationship. By creating a board game without rules and with chess pieces made of the colours and shapes of the Chinese and American national flags, the play and the result of the game depended entirely on random interference of the audiences/players. The hole in the middle of the board, indicating the political and psychological border, enhanced the sense of crisis and limitation.

At the same time, the idea of *shanzhai* in relation to the comparison of the two dreams and mentalities emerged during our discussions. Like other Rust Belt localities, such as Pittsburgh and Cleveland, Buffalo was the city that once witnessed the realisation of the American Dream. At the dawn of the twentieth century, Buffalo was called the City of Light due to the widespread electric lighting. It was also part of the automobile revolution; the development of the industry drew a great number of immigrants from industrial European countries like Germany and Poland to pursue their American Dream here. With the opening of the St. Lawrence Seaway in 1957, which cut the city off from valuable trade routes, deindustrialisation, and the nationwide trend of suburbanisation, the city's economy began to deteriorate. In the 1950s, Buffalo saw its population decline as heavy industries shut down and people left for the suburbs or other cities. In the glorious architectures in downtown Buffalo, the massive silo cities on the

2.6.1 **Teng Teng, Zhou Quan and Liz Black 'I Am in the Dark Here'**

2.6.2 **Detail of 'I Am in the Dark Here'**

2.6.3 Lau Cherry Yung Yung, Morgan Layne 'Double Rule'

south shore of Lake Erie, and the monumental power plant at the Niagara Falls, I saw the manifestations of the Dream come true. However, the reality of the Rust Belt city in decline also reminded me, a Chinese person whose compatriots are copying the same dream, of this scenario: what it could look like *after* the 'dream' was realised. That's what makes Buffalo a fascinating place for artists: what is the aftermath of the Dream come

2.6.4 Zheng Xinyu, Caroline Doherty 'Rule Rules Ruled Ruler'

true? Or, to put it in another way, what happens if the Dream evaporates? The big question mark soon became the centre of most of the students' work. Naturally, the possibility of using *shanzhai* as a major curatorial strategy had been undertaken by us. It gave us the opportunity to ponder the implications of the *shanzhai* mentality for the art world and its cultural politics.

2.6.5 Over Looking Down Town Buffalo form the City Hall Tower

The work that directly used this tactic was created by Fu Xin, Lin Ru and Alyssa Crane. Although it did not borrow the idea of the 'dreams' directly, the work *Smuggling Fu Xin into the U.S.* was a perfect example of how the idea of *shanzhai* was used in our programme. The concept of this piece was a result of a given circumstance. A Chinese participant of our group, Fu Xin, could not make it to Buffalo with the rest of his fellow students. He had been rejected, not once, but twice on his Visa request at the U.S. Beijing embassy. The reason of admittance rejection of this graduate student from a prestige university with an academic travel purpose remained a mystery; it is a customary rule of the embassy that the specific reason of denial is not to be given. However, this inspired the artists to create a work to fabricate a virtual Fu Xin in the exhibition by recreating a life-sized scanned image, along with photocopies of his Visa application files, rejection notices from the embassy, his passport pages, finger scanning images and other identity certification documents. Here, a *shanzhai* person was represented in a physical environment by technological means, while the concept of *shanzhai* was shifted to a critical context: a constrained artistic manoeuvre to resist a supreme power, at the same time challenging geographical and political boundaries. It also questioned the parochialism of the 'national dream' mentality.

All of the above examples engaged with the history and cultural background of the exhibiting site and the concept of the *shanzhai*. It requires close collaboration between artists and the exhibition site to develop a complex exhibiting space, not to mention studying the history and cultural background of the location. Nevertheless, I still consider such engagement essential to our exchange programme, because it can initiate debates in art teaching, art making, and curatorial practice, which constitute the most dynamic and exciting aspects of contemporary art education. Here, a possible curatorial strategy has emerged: the collaboration works of the artists interacted with the exhibition site on multiple levels; they made the audiences ponder on the common problems of the protection of both natural and human environment as well as on the interplay of every aspect of the two 'Dreams', which would associate, enlighten, and resonate with each other.

From this perspective, it corresponds to concepts such as 'dialogic' and 'intertextuality' in post-structuralist philosophy. Mikhail Bakhtin's dialogism suggests a continuous dialogue with other works of literature and other authors – and his examination of the multiple meanings, or 'heteroglossia', in each text (especially novels) and in each word. Building on Bakhtin, Julia Kristeva argues that 'the notion of intertextuality replaces the notion of "intersubjectivity" when we realise that meaning is not transferred directly from writer to reader but instead is mediated through, or filtered by, "codes" imparted to the writer and reader by other texts'.[1] With the help of the interaction between the art work and the exhibiting space, the intertextuality of fantasising the realisation of the Chinese Dream, or the *shanzhai* American Dream, activates a dynamic relationship: a psychological perception and a dialogue could be found between the audience and the art work, hence the audience has become a part of the exhibition. This, in turn, illustrates the central curatorial strategy of contemporary art: to create various kinds of interactions between space, images, and objects. Some images and objects bring back memories of the past, others interact with the present. Mixing and overlapping, they generate confusion in our historical perception, as if past and present were staged simultaneously by means of transforming objects into a visual medium.

The curatorial practice cited for these field notes was carried out revolving the axis of the *shanzhai* 'dreams', to explore how the boundaries between copycat and original were settled and unsettled during the process of subjective identification, amidst an attempt to capture the artistic agency wavering from reality and imagination. All the projects created by the

1 Julia Kristeva, *Desire in Language: A Semiotic Approach to Literature and Art* (New York: Columbia University Press, 1980), 69.

students revealed the political, cultural, and economical factors informing the national 'dreams', problematising the concept of *shanzhai*, and fostering a network of mutual understanding, ultimately summoning up criticality with the tactics of subjectivisation and defamiliarisation. It formed a personal interpretation that challenged the national narrative; the concept of the 'Chinese Dream' became problematic.

To summarise, our curatorial strategy throughout this programme was a manoeuvre of time and space, at the same time using the concept of *shanzhai* as one of the running threads of imagination and inspiration. All the projects reviewed the regional history, geographical politics and contemporary social dynamics, embedding them in a broader cultural/economic context to advance critical speculation: the political, economical and globalised framework of the 'Chinese Dream' ideology.

Video clips

A speech competition held in an elementary school during the promotion campaign of the 'Chinese Dream':
http://baidu.ku6.com/watch/7388180887690985814.html?page=videoMultiNeed

Official promotion cartoon of 'Chinese Dream' issued by the Publicity Department of the Communist Party of China:
http://www.iqiyi.com/w_19rrssrc3x.html

Bibliography

Friedman, Thomas. 'Xi Jinping and the Chinese Dream', *The Economist*, May 4, 2013. https://www.economist.com/leaders/2013/05/04/xi-jinping-and-the-chinese-dream.

Kristeva, Julia. *Desire in Language: A Semiotic Approach to Literature and Art* (New York: Columbia University Press, 1980).

Mars, Neville, and Adrian Hornsby. T*he Chinese Dream: A Society Under Construction* (Rotterdam: 010 Publishers, 2008).

Wang, Helen. *The Chinese Dream: The Rise of the World's Largest Middle Class and What It Means to You* (CreateSpace Independent Publishing Platform, 2012).

Wu, Hung. *Making History: Wu Hung on Contemporary Art* (Hong Kong: Blue Kingfisher, 2009).

Wu, Hung. *Wu Hung on Contemporary Chinese Artists* (Hong Kong: Blue Kingfisher, 2009).

Xi, Jinping. *Xi Jinping: The Governance of China* (Shanghai: Shanghai Press, 2015).

About the Author

Feng Fan is an artist, Assistant Professor at Academy of Arts & Design, Tsinghua University, faculty of The Institute of Contemporary Art, Tsinghua University. Feng Fan is also a Doctor of Arts, who has exhibited work across China, and in the U.S. and Japan. Feng's work is in several public collections and he has produced a number of public art commissions around China.

2.7 'Banal Creativity'

What Does It Mean to Be Creative for Hunan TV Practitioners?

Arjen Nauta

Abstract

Hunan Satellite Television has established itself in the past decade as the most innovative and popular regional TV station. Based on fieldwork at HSTV in Changsha, I show how television makers make no clear binary distinction between creative and non-creative, or 'original' and 'imitation'. This conceptual space, which challenges the dichotomy between creativity and imitation, is where 'banal creativity' takes place. The dichotomy between creativity, and related concepts such as originality on the one hand, and copying, mimicry, and imitation on the other, does not take the middle positions into consideration and moreover establishes a misleading idea of creativity and originality as 'ex nihilo'.

Keywords: Hunan Satellite TV, fieldwork, *shanzhai*, banal creativity, Chinese television

'Hunan TV is more creative than other TV stations'. I have heard this statement repeatedly when asking my interviewees what makes Hunan Satellite Television (HSTV) different from other TV stations. Its rise has been remarkable: largely since the turn of the century, HSTV has been the most popular provincial television station in China, despite its disparity in income, resources, and material compared to satellite television from richer provinces, such as Zhejiang or Shanghai.[1] I set out to do fieldwork in Changsha, the capital of Hunan and basis of HSTV, for a period of ten months (Sep 2015-Jun 2016) to

1 The Chinese television system in tier-based. China Central Television (CCTV) is the only national broadcaster. Since 1997, all provincial stations have one satellite channel that is broadcast nationally. Hence, HSTV can be viewed all over China.

Kloet, Jeroen de, Chow Yiu Fai, and Lena Scheen (eds), *Boredom,* Shanzhai, *and Digitisation in the Time of Creative China*. Taylor & Francis Group, 2019
DOI: 10.5117/9789462984745_CH2.7

find out, in the first instance, how it is possible that a television station from a poorer, non-coastal province could obtain the top spot among its more affluent competitors. And if the first answer of all my interviewees (who are or have all been HSTV employees) is that HSTV is more creative, then what does that mean? What does this creativity entail, and how is it understood? What does creativity mean for HSTV employees in their daily job? And how do their conceptions of creativity relate to global discourses? As Eitan Wilf argues earlier in this book, global conceptions of creativity are contingent on a Western-Modern-Romantic discourse. So how do Chinese television makers interpret this concept and how does it speak back? These field notes are based on 27 semi-structured interviews with television makers (producers, directors, scriptwriters, etc.) from HSTV, conducted in Changsha.

In 1997, the Radio and Television Bureau of Hunan Province (RTBHP) launched Hunan Satellite TV. Ever since its inception, HSTV has focused on producing alternative content, distinguishing itself from CCTV and other satellite stations, by focusing on interactive entertainment television.[2] It was the first station to productise localised versions of successful foreign formats, and still relies heavily on this strategy. However, in the week of 12-18 March, HSTV broadcasted only one foreign show: the fourth season of the localised version of the Korean show *I am a singer*, now rebranded as *Singer*.[3] The main reason for this absence of foreign formats is a series of regulatory edicts issued since 2013 by the State Administration of Press, Publication, Radio, Film, and Television (SAPPRFT), the regulatory body for Chinese television, restricting foreign content. From 2014 onward, television stations have only been allowed to broadcast one foreign format per year.[4] However, HSTV has found several ways to circumvent these restrictions.

When my interviewees state that HSTV is more creative, I always ask why. And they always remain silent for a few moments. Then, the answer is twofold. First, they emphasise, creativity is institutionally stimulated. HSTV's motto is 'create or die' and brands itself as a creative hub for its employees

2 Yong Zhong, 'Hunan Satellite Television over China', *Journal of international communication* 16 (2010): 41-57; Michael Keane, 'As a Hundred Television Formats Bloom, a Thousand Television Stations Contend', *Journal of contemporary China* 11.30 (2002): 5-16.

3 Yajia He, '"I am a Singer" Renamed "Singer" is More Competitive than any Previous Season ['Wo shi geshou' gengming 'geshou' jingyan que bi ciqian renhe yi ji dou jilie]', *People's Daily*, accessed December 16, 2016. http://ent.people.com.cn/n1/2016/1215/c1012-28951304.html.

4 See reports by *People's Daily* (Chinese) or *Hollywood Reporter*, such as: Clifford Coonan, 'Chinese Regulators Restrict Imports of TV Formats', *Hollywood Reporter*, accessed March 5, 2017, http://www.hollywood reporter.com/news/chinese-regulators-restrict-imports-tv-649704; Patrick Brzeski, 'China Ccracks Down on International TV Formats', *Hollywood Reporter*, accessed March 5, 2017, http://www.hollywoodrep orter.com/news/china-cracks-down-international-formats-904345.

and as reason for its success. Most HSTV employees I spoke to emphasise their excellent in-house training academy, where aspiring television makers learn how to create successful programmes. The organisational structure itself is geared towards stimulating creativity, they contend. There is no quagmire of management layers between the production teams and the leadership, and any employee – from shopping channel hosts to parking attendants – can come up with ideas for new programmes that will seriously be considered. Of the 30 production teams, only ten can make a programme at the same time; the other teams try to force their way in by developing new ideas and formats. If a production team has not produced a programme that is selected for broadcasting for two years, the team is dissolved – literally 'create or die'. However, when the term creativity is so institutionalised and even sloganised, perhaps it should primarily be taken as a meta-narrative that configures how HSTV and its employees describe themselves and their creative work.

The second answer points to HSTV's development in the early 2000s. This was a time, they claim, when HSTV was really willing to experiment, and move into new areas in order to strengthen its position vis-à-vis its competitors. At this time, HSTV was one of the first Chinese television stations to produce localised versions of foreign television formats. Localisations of popular foreign dramas and reality shows, such as Taiwanese *Meteor Garden* and *American Idols* were highly successful, whereas HSTV's version of *Big Brother* failed to catch on.[5] Most of these programmes were unlicensed, possible due to lax copyright enforcement in China. Production teams 'cloned' foreign formats, but by making minimal changes and giving them a different name, claimed them as originals.

'Banal Creativity'

Asking further, beyond institutionally advocated meta-narratives of creativity, into the meaning of creativity in the daily work of my informants, claims of creativity became more ambiguous. While on a meta-level, interviewees considered many of these programmes to be copies, and not really examples of the 'creative spirit' of HSTV, they simultaneously claimed that on a micro-level, in their daily job, the localisation process in itself was highly creative. Without a license, production teams lacked advice and support from the copyright owner, and thus it fell completely upon them to remake the show for a Chinese audience. But even when licenced, characters, symbols,

5 Keane, *The Chinese Television Industry* (London: BFI Palgrave, 2015), 86.

and narratives need to be omitted and/or adjusted in such a way that the programme resonates with Chinese audiences, a process that, according to my interviewees, is creative in itself.

And whereas in some cases the links between a localised version of a programme and a global format is quite clear, in other cases it becomes more complicated. One way in which HSTV stimulates the development of original content is by letting production teams watch foreign shows as inspiration to create new shows by combining elements from different shows. An interesting example is *X-change*, according to its producer loosely based on the UK-format *Wife Swap*. In this programme, two families, usually from different social classes, swap wives/mothers for to weeks. The Chinese spin-off diverges in three important ways. Firstly, instead of the wife/mother, the child is being swapped; secondly, the swap lasts just one week (one month in later seasons); and thirdly, the programme focuses more on urban-rural than class divisions; one child comes from a poor countryside village, while the other grows up in a rich urban family. The programme has such a different outlook from the British format that one production cast member who I interviewed was not even aware of its British origins.

While this idea of creativity might sound banal to Western audiences, I find that, based on my research, television makers make no clear binary distinction between creative and non-creative, or 'original' and 'imitation'. One executive producer told me that he considered *X-change* to be twenty per cent imitation, and 80 per cent original. While no other interviewee explained it in such a clear way, others were unequivocally offended by terms such as 'cloning' or 'copying' when referring to their own work, which suggest that there is nothing creative or original about their endeavours. According to them, no matter whether a programme is based on a foreign format, the process of localisation entails extensive adjustments and omissions, character reformations, etc. – a creative process in their view, perhaps second-degree creativity, but which must at least be recognized as such. This conceptual space, which challenges the dichotomy between creativity and imitation, is where 'banal creativity' takes place. However, can we take these statements at face value?

I argue that we should. Interestingly, the opinions of my interviewees do not challenge Western conceptions of creativity, nor the all-pervasive global glorification of the 'creative industries'.[6] During the conversations I had with them, my interviewees never challenged Western-Modern-Romantic conceptions of creativity as being 'unfit' or 'unsuitable' for China. When

6 Li Wuwei, *How Creativity Is Changing China* (London: Bloomsbury Academic, 2011); Richard Florida, *The Rise of the Creative Class* (New York: Basic Books, 2003).

talking about creativity on a meta-level, they pointed to the development of new formats, very much in line with the standards of global format trade. However, their claims of what I term banal creativity do not challenge this discourse, but rather extend the conceptual space where creativity can take place and should be recognised as such. Banal creativity thus refers to the opening-up of the domain of the creative, recognising the daily creativities beyond the grand claims of originality and uniqueness vis-à-vis imitation and copying. Moreover, this wider conceptual space where banal creativity can take place is not just limited to the television industries.

Shanzhai and Banal Creativity

My findings are closely related to the emergence of *shanzhai* culture since 2008. While resistant to clear definitions, *shanzhai* things typically emerged from grass-roots interventions in production beyond the control of the authorities.[7] *Shanzhai* products are 'almost the same, but not quite'. Lin Zhang and Anthony Fung analyse how a *shanzhai* Spring Festival Gala emerged as a bottom-up challenge to the political and economic monopoly of CCTV's annual Spring Festival gala.[8] These grass-roots creative interventions are disruptive of systems of domination, such as intellectual property (IP), since 'it creates gaps in the systems of knowledge through which authority is exercised'.[9] At the same time, many *shanzhai* products, such as phones and paintings, are deeply embedded in, and reliant on, the structures of global and national capitalist economic and consumer culture – localisations of global products for more practical local application.

Although the copying of formats has been marked *shanzhai* as well, and HSTV's popularity is often perceived as a bottom-up challenge to the domination of CCTV I do not wish to conflate the practice of appropriating formats with the term *shanzhai*. What matters is that *shanzhai* fills a similar conceptual space on the line between creativity and imitation.[10] *Shanzhai* products challenge this dichotomy in similar ways. Creativity in *shanzhai*

7 Andrew Chubb, 'China's *Shanzhai* Culture: "Grabism" and the Politics of Hybridity', *Journal of Contemporary China* 24.92 (2015): 260-79.

8 Lin Zhang and Anthony Fung, 'The Myth of "*Shanzhai*" Culture and the Paradox of Digital Democracy in China', *Inter-Asia cultural studies* 14.3 (2013): 401-16.

9 Chubb, 'China's *Shanzhai* Culture', 263.

10 Unlicenced reproduction of foreign TV formats has been labelled *shanzhai*. For example, see *People's Daily*, '中国电视节目频谱"山寨"国外"他山之石"总灵光? [Zhongguo dianshi jiemu pinpu '*shanzhai*' guowai 'ta shan zhi shi' zong lingguang]', accessed September 18, 2016.

is loosened from its correlations of authenticity and originality. By making creative interventions in the production chain, *shanzhai* products are copies but not quite. In the same way, television makers regard the localisation of foreign formats as a creative process. While these conceptions do not challenge the way in which the discourse of creativity revolves around Western-Modern-Romantic conceptions of authenticity and originality, they do destabilise its dichotomous position vis-à-vis possible antonyms such as imitation or mimicry and extends the conceptual space for creativity.

My argument is therefore not a Chinese challenge to the Western-modern-romantic discourse of creativity, pointing to some self-Orientalising Chinese otherness. Rather, the results from my fieldwork, epitomised in the concept of banal creativity, calls for a conceptual extension of the recognition of creativity. The dichotomy between creativity, and related concepts such as originality on the one hand, and copying, mimicry, and imitation on the other, does not take the middle positions into consideration and moreover establishes a misleading idea of creativity and originality as 'ex nihilo'. The concept does however serve to hold a mirror to the ideas of creativity featuring in Western countries. Closing off with an example from television, why should *The Voice*, *The X-Factor*, *Idols*, and the *Got Talent* franchises be regarded as original formats? They have so many elements in common, that their claims of originality as compared to a show like *X-change* seem negligible.

This research is supported by a consolidator grant from the European Research Council (ERC-2013-CoG 616882-ChinaCreative).

Bibliography

Brzeski, Patrick. 'China Cracks Down on International TV Formats', *Hollywood Reporter*. Accessed March 5, 2017. http://www.hollywoodreporter.com/news/china-cracks-down-international-formats-904345.

Chubb, Andrew. 'China's *Shanzhai* Culture: "Grabism" and the Politics of Hybridity', *Journal of Contemporary China* 24.92 (2015): 260-79.

Coonan, Clifford. 'Chinese Regulators Restrict Imports of TV formats', *Hollywood Reporter*. Accessed March 5, 2017. http://www.hollywood reporter.com/news/chinese-regulators-restrict-imports-tv-649704.

Florida, Richard. *The Rise of the Creative Class* (New York: Basic Books, 2003).

He, Yajia. '"I am a Singer" Renamed "Singer" is More Competitive than any Previous Season [Wo shi geshou' gengming 'geshou' jingyan que bi ciqian renhe yi ji dou

jilie]', *People's Daily*. Accessed December 16, 2016. http://ent.people.com.cn/n1/2016/1215/c1012-28951304.html.
Keane, Michael. *The Chinese Television Industry* (London: BFI Palgrave, 2015).
—. 'As a Hundred Television Formats Bloom, a Thousand Television Stations Contend', *Journal of contemporary China* 11.30 (2002): 5-16.
Li, Wuwei. *How Creativity Is Changing China* (London: Bloomsbury Academic, 2011).
People's Daily. '中国电视节目频谱"山寨"国外"他山之石"总灵光? [Zhongguo dianshi jiemu pinpu '*shanzhai*' guowai 'ta shan zhi shi' zong lingguang]'. Accessed September 18, 2016.
Wilf, Eitan. 'Semiotic Dimensions of Creativity', *Annual Review of Anthropology* 43 (2014): 397-412.
Zhang, Lin and Anthony Fung. 'The Myth of "*Shanzhai*" Culture and the Paradox of Digital Democracy in China', *Inter-Asia cultural studies* 14.3 (2013): 401-16.
Zhong, Yong. 'Hunan Satellite Television over China', *Journal of International Communication* 16 (2010): 41-57.

About the Author

Arjen Nauta is a doctoral researcher at the Department of Media Studies, University of Amsterdam. His project on Hunan Satellite Television is part of the ERC-funded 'China Creative'-project, led by Prof. Jeroen de Kloet. He obtained a Research Master's degree in Religion and Culture and two Bachelor's degrees (history and religious anthropology) at the University of Groningen. In addition, he received a master's degree in China and Asia-Pacific Studies from National Sun Yat-sen University in Taiwan. From August 2015 until June 2017, Arjen resided in Changsha, Hunan, to do his fieldwork at HSTV.

2.8 Two sides of SZ (*Shanzhai* & Shenzhen)

Lo Yin Shan, Dai Dai, and Deng Chunru

Interviewer: Lo Yin Shan
DD: Dai Dai (Chinese artist/ curator of Wen Bo Gong, Shenzhen)
DCR: Deng Chunru (Chinese artist/ founder of New-Who Art Museum, Shenzhen)

This dialogue was conducted in Chinese and translated into English by Sonia Wong.

Q1: When did you move to Shenzhen? Why did you choose Shenzhen? How is Shenzhen different from other first or second-tier Chinese cities in terms of its way of life and culture? What aspects would you describe as uniquely Shenzhen (for example her geographical location at the border of the country)?

DD: I moved to Shenzhen in 2008, one year after I graduated from university. It was by chance that I came to Shenzhen, not something I have planned for. I didn't consider moving here because back then I was just starting my painting career, and I felt that Shenzhen didn't really have an art scene. But, for me, Shenzhen has greater tolerance in terms of lifestyle and arts and culture compared to other cities. As a young migrant city, it welcomes the ecology of multi-diversities which makes the city lively and full of possibilities.

DCR: I was born in the old Hakka village Aohu in Shenzhen in 1977; all my ancestors lived here, so it wasn't really a choice. I haven't spent much time in other cities so I couldn't tell how they are different from personal experience, but Shenzhen is undoubtedly unique: Shenzhen's geographical location characterised her experimental nature and cultural personality, which permeated into all aspects of life. In a way, all cultures are geographically fated.

Kloet, Jeroen de, Chow Yiu Fai, and Lena Scheen (eds), *Boredom,* Shanzhai, *and Digitisation in the Time of Creative China*. Taylor & Francis Group, 2019
DOI: 10.5117/9789462984745_CH2.8

2.8.1 The founder of Xiaomi, Lei Jun claimed Steve Jobs as his model and icon

As to how Shenzhen is unique, I hope my answer is not too clichéd. The character of any border city is largely determined by whom she is bordering – if Shenzhen had been next to North Korea instead, Shenzhen would not be like we know it now. So, we cannot avoid mentioning Hong Kong, because Shenzhen would not be if Hong Kong was not Hong Kong.

Q2: Can you talk a bit about how Wen Bo Gong (文博宫) / New-Who Art Museum (鳌湖美术馆) and the art village came to be, your relationship to these places and some interesting projects here which you have worked on? Also, maybe you could share your experience and stories of artistic practice of everyday life in Shenzhen?

DD: I first came to Wen Bo Gong (Culture Exhibition Palace) in 2012, when it was originally a government-planned 'food-court city', but later on it was acquired by some private enterprise group. Because the land was designated for the function of cultural tourism, so that company, after

2.8.2 The Chinese calligraphy as a façade for the official Apple Store in Hangzhou

some research, decided to model it after the Panjiayuan Antique Market in Beijing. In the earlier stages of development, they introduced a lot of dealers specialising in all sorts of antiques and curios, setting up stores for selling rare rocks, antique books and paintings, and so on. Later , in order to distinguish itself from Xinxiu Antique City that is also in Shenzhen, they set up an Intangible Cultural Heritage exhibition area, and the whole new Cultural Industry Park gradually takes shape. I am in charge of the art museums in the district and Xiangshan Museum (祥山艺术馆) is the only one that is ran independently by our company. At Xiangshan I have co-curated over twenty exhibitions, and the most memorable one is the 'Star is Rising – The Second National Art College Graduates Biennial Promotion Exhibition' in 2014. The exhibition was headed by curator Wang Lin and included works by students from the nine major art colleges. We have invited ten renowned artists and critics to form a team, and they selected 31 outstanding young artists for the showcase and presented ten prizes as recognition of their work.

Unlike Beijing, Shenzhen does not have what we call 'professional artists'; most artists or curators here all have different kind of jobs, I think it is really the passion for art that keeps them going. When I first came to Shenzhen in 2012, my job took up so much of my time that I could hardly paint, so everyday after work I would go to bed at seven or 8pm, and get up at 2am to paint until the next day, and go straight to work at 7am. This routine lasted over six months. Seven years in Shenzhen, I have always had a job, but I've never given up on my art.

DCR: Around the time of Chinese New Year in 2012, my mother expressed her wish for me and my family moving back from the new downtown of Shenzhen to the old Hakka village Aohu, so we can spend more time with her. My mother has her own place, so even if I move back, I won't be staying with her, but we live close enough to each other. After I've discussed this idea with my wife, and got her consent, we moved back – which is nice because now my home is just a ten-minute walk from my studio. Even though my wife had a hard time adjusting to the life in a Guangdong rural village, but the prospect of more spacious living and not having to fight over parking space gives her enough motivation to try [...] so we agreed on moving back. This is the beginnings of New-Who Art Museum – my homecoming.

When I first moved back, the old Aohu Village (a hundred-year-old Hakka village with forts) is populated by people from the very bottom of society: it's disorganised and chaotic, with all sorts of danger lurking underneath. To make things worse, there were a lot of illegal factories, polluting the village with toxic gas and poisonous liquid waste. Theft was frequent. My wife and I simply could not adjust to all these, and for me, my heart was filled with sadness and a profound sense of loss.

When I walked among the ruins of the old village, places that recalled memories from my childhood, juxtaposing that happiness from the past with its wretched present state gave me immense psychological pressure. I was overwhelmed by a sense of nostalgia, I longed for my 'homecoming', but I felt cut off at the same time.

My action was an answer to this emotional burden I felt. I started inviting my friends from the art circle to visit the village in the name of traditional Hakka architecture touring, trying to 'convince' them how old village house with courtyard would make great studio spaces. Thanks to the high real estate price and rent of Shenzhen, you didn't need to be a great salesman, because the rent at the village was so low that only several hundred RMB a month for a huge 'yuan' complex. As more and more artists move in, most of the illegal factories were displaced out of the village, and artists have the most amazing ideas and abilities to transform the old houses. In short, with all the efforts put in, I felt the burden being gradually lifted, and new projects keep flying in that I can hardly find time to rest. I am truly grateful for everything everyone has contributed to this place.

My relationship with Aohu is about me and my homeland, my past, present, and future.

On meaningful projects: for all the experiments and practices we have done at Aohu, there are a lot of interesting projects. For example, there is a 'village store' project, which my wife and I (she is also a curator) are working on, we try to incorporate art into the daily set-up of small village stores, while trying to keep everything else untouched, it's a whole new cultural processing.

With artists living in this community, constant interaction is possible, and with this constant contact there comes inspiration and creation, which makes artists who are not part of this community rather envious. I envision all these to be parts of a bigger, more fascinating and demanding project of Aohu – the New-Who Art Village, the New-Who Art Museum – they are the most interesting and meaningful artistic endeavours of mine.

I would describe my daily life in Shenzhen as pretty ideal by normal standard – my personal life is happily married with my work, I can make a living doing what I love and is capable of, I have a group of close friends who share my interests and visions, we meet up as often as we like, my child studies nearby, I can get together with my family at the dinner table everyday. My visions are all incorporated into my life and work, so I would not further elaborate. Some people might think I am naïve, some other could just be jealous.

Q3: According to Baidu and Wikipedia, there are several possible origins for the notion of shanzhai*: a)* Shanzhai *factory and product originated from the Cantonese Hong Kong in the 1960s to 1970s (e.g. Li Kasing's plastic flower production), then 'migrating' to Shenzhen at the end of the 1970s. b) Shenzhen was the heart of fake electronic appliances during the 1990s, but manufacturers could only be marked 'made in SZ', so 'Shenzhen product' gradually became '*shanzhai *product'. c) Dafen Oil Painting Village – in the early 1990s a group of about twenty artists under the leadership of the painter and businessman Huang Jiang, who came to Dafen in 1989, took up residence in this town. d) The Chinese character of* shanzhai*, literally means 'mountain stronghold' or a fenced place in the forest, refers to the mountain stockades of regional warlords or bandits, far away from official control. Historically, the term is sometimes used as a metaphor to describe bandits who oppose and evade the corrupted authority to perform deeds they see as justified. One example of*

2.8.3 The copycats of Zaha Hadid built much faster than the original architecture

such bandits is the story of Chinese classic Water Margin (水滸傳, *alternative title* Outlaws of the Marsh). *Can you explain how you understand* shanzhai, *its origin, and its relations to Shenzhen?*

DD: *Shanzhai* for most of us is a sarcastic term. The emergence of *shanzhai* culture should coincide with China's overall development after the 'open-door' policy; I imagine it's a phase every developing country would go through.

DCR: *Shanzhai* is not only a mere concept, but has a whole world within! There are countless forms and categories and levels of s*hanzhai*, and it is this richness that draws you, the academics, into studying it. I don't think we should be preoccupied with locating a specific 'origin', but to view it as a necessary process and phase in the natural development of things.

Shenzhen dares trying whatever that is new and curious, because we don't have the burden of a long history, that's why the evolution process can be speeded up. With such velocity, I assume Shenzhen would be the first to grow out of the s*hanzhai* phase.

Q4: It seems that for some, shanzhai *is an overtly negative term, but this phrase and concept is actually loaded with meaning, and can be interpreted in many different ways, pointing not only to how China has developed as a creative nation, from 'made in China' to 'created in China', but also the development of* 'shanzhai *aesthetics', 'second degree of creation' or 'spoofing culture' and so on. What do you think of it? And could you give us some cases from daily life or within the cultural creative industry as examples?*
DD: I think the progression from 'made in China' to 'created in China' is simply a process of modern development. *Shanzhai* culture is a phenomenon organically generated from the IT sector and popular internet users, representing an unorthodox grass-root subculture, which is in itself hugely flawed – their lack of critical thinking in choices and identification, making it closer to parody or profanity than genuine creativity.
DCR: *Shanzhai* in itself is a display of the courage of self-ridicule, which makes it somewhat admirable and respectable. We can see how China as a newcomer in international businesses tries to use s*hanzhai* as an 'undermined' strategy to make her presence known, and that it works. On the other hand, by standing on the shoulders of giants, we can see how we are able to produce something uniquely ours in this process of learning and copying, sometimes the quality and ingenuity of s*hanzhai* products even surpass that of the original. I think this competition is a force of progress that we would benefit from.

As much as I am not willing to admit, WeChat is a s*hanzhai* collective of existing foreign mobile apps, but from copying and modifying, this mega app is now the most convenient and powerful app available. A lot of Hong Kong people look down on WeChat, and are unwilling to use it, I think they don't know what they've missed. Now we don't even need to carry our wallets around – your daily matters can all be sorted out with a single app – apps like WhatsApp or Line are really elementary in comparison. I sincerely urge friends from Hong Kong, especially those who have frequent contact with China, to have a good grasp of WeChat. WeChat has long ceased to be '*shanzhaing*'; it is a unique creation by itself.

Q5: How do you comprehend the historical development of Dafen Oil Painting Village since its founding in 1989 (according to sources online), for example, from spontaneous organisation and organic development to government intervention of establishing Dafen as 'National Cultural Industry Demonstration Base'? Do you know residents/craftsmen/artists from Dafen? Any interesting stories to share? How do you think the 'atmosphere' of the place would affect (pro and con) those who live/work/grow up there?

DD: The emergence of Dafen, for me, is something incredible and unthinkable. We can also see how it reflects the society's general lack in aesthetics and arts education, Dafen has nothing to do with art. I know a lot of people from Dafen, but we don't have much artistic interaction, as they are only trying to make a living. This atmosphere is quite unhealthy and definitely harmful to those who truly love art, like our Fine Arts College Entrance Examinations.

DCR: Dafen Oil Painting Village is a miniature of China's manufacturing industry, as 'cultural manufacture' is also a part of the whole industry.

I know a lot of residents/craftsmen/artists from Dafen, some with whom I am quite close. Shenzhen is not a big city, it's natural that cultural practitioners and artists would know each other. The people I know from Dafen come from a wide spectrum: artists who insisted on creating original works and have their own galleries (they live there because of the flow of people and number of visitors), owners of workshops that produce replicas (who ship cargos after cargos of replicas to Europe every year), middle-aged art lovers who open cafes in the village, replica painters who make a living by taking small orders, or even villagers who rent out their places. We would meet up at times (usually over north-eastern Chinese cuisine), to hang out and chat, nothing special.

Dafen has a strong sense of a collective, which makes it easy for individuals to feel powerful and important being part of this collective, but once they are uprooted from this community, it's not easy to find an anchor point upon which they can build self-worth.

About the Authors

Lo Yin Shan graduated with a diploma in design from the Hong Kong Polytechnic University, and is currently a feature reporter/editor and an amateur artist/photographer. She lived in Beijing from 2008 to 2017 as a 'teleporter'/columnist between two cities. Author of *Driving Lantau: A Whisper of an Island*; *Ten Notes on Northern Drifting*; editor of *The Box Book: Beauty and the Beast of Hong Kong Culture.*

Dai Dai is an artist and curator. He was born in Anhui in 1987 and graduated at Xi'an Academy of Fine Arts majoring in Chinese and Western History of Fine Arts and Oil Painting. He is currently living and working in Shenzhen. Since 2008, Dai Dai was appointed as art curator of Contemporary Gallery, and Acting Executive at Private Chinese Museum. Currently he is engaging in Contemporary Arts Exhibition works.

2.8.4 A Chinese hymn celebrates 'the power of Shanzhai'

Deng Chunru, born in Guanlan, Shenzhen in 1977, graduated from Guangzhou Academy of Fine Arts in 2001, now living and working in Shenzhen. He also engages himself in arts creations, experiments and organisations, such as the New Who's Voice, Qiming Restart-Culture Rebuild for Edge Community art exhibitions in 2013, as well as the performance project New Who Live and group exhibition Aohu Discovery in 2014. His solo exhibitions include 'The expression of the time in the creation' Hongkong UMA-G Gallery (2005), 'Flowers Depths' Shenzhen Art Museum (2007), 'A river of ordinary life' Shenzhen ZaiGallery (2013), Container – Shenzhen Art Gallery (2013). His paintings are collected widely by art museums. The author of 'My Little world – Manuscripts of Deng Chunru' and 'Deng Chunru Oil Paintings'. Member of Guangdong Province Art Association, Vice-president of Shenzhen Young Artists Association, Council of Shenzhen Art Association, Member of Oil Painting Art Association, the initiator of 'Qiming Restart' project, principal of Aohu art museum, Practitioner of rural community culture rebuild.

Section 3

Digitisation

ESSAY

3.1 Creativity, Affordances, and Chinese Traditional Culture

Michael Keane

Abstract
In this chapter, Michael Keane examines the traditional roots of Chinese creativity. He uses the concepts of *creatio in situ* and affordances to show that Chinese creativity is both contextual and relational. Keane examines the connections between Confucian and Daoist approaches to creativity, nature, and the market and provides a personal view of the trials and tribulations of creativity in China over the past two decades. Finally, he shows how the concept of affordances provides a new way of understanding Chinese creativity. Persons, things, and discourse take on many appearances as time goes by. The manifestation of *shanzhai* culture in this sense demonstrates such an affordance.

Keywords: digital China, creative industries, Chinese philosophy, creative economy, *shanzhai*, Confucianism

Introduction: Innovation is the Soul of a Nation's Progress

Never has the need to innovate been so pressing. In a time of volatile markets, natural disasters, falling commodity prices, rising national debt, and decreasing returns on manufacturing, innovation is evoked as the mantra of national and regional rejuvenation. While research and advocacy on National Innovation Systems has been around since the early 1980s[1] and a

1 Sometimes NSI or NIS. For a discussion of origins see Lundvall et al. (2007) and for its application in creative sectors, Mitchell et al. (2003).

Kloet, Jeroen de, Chow Yiu Fai, and Lena Scheen (eds), *Boredom,* Shanzhai, *and Digitisation in the Time of Creative China*. Taylor & Francis Group, 2019
DOI: 10.5117/9789462984745_CH3.1

Global Innovation Index has existed since 2006,[2] the pace of technological change is accelerating, leaving many nations behind, leading to speculation about the 'Industries of the Future'.[3] Americans *must* innovate because the Chinese are coming to take away their jobs; Britain *must* innovate because European talent is leaving its shores; Australia *must* launch an 'ideas boom' to reduce reliance on its resources economy.[4] China *must* have more innovation because its export-based economy is floundering and its population ageing.

Very few governments today are immune from the type of techno-utopianism spruiked by Silicon Valley prophets of progress and TED.com problem-solvers, the latter variety emblematic of what Benjamin Bratten calls 'middlebrow megachurch infotainment'.[5] Technological innovation will help resolve global problems, from climate change to cancer. Crowdsourcing, crowdfunding, the sharing economy, block chain technology, and self-organising networks will break the dominance of governments and multinational corporations, kids will learn how to code in kindergarten, and paper currency will be a museum exhibit.[6]

Then there is the cloud, a distributed system of computer resources, millions of servers, hard drives, routers, fibre-optic cables, and networks that connect masses of people while tracking their online lives.[7] The Chinese government believes this digital cloud, which also crosses into sovereign territories, will transform ordinary people into entrepreneurs who will innovate, invoking a new wave of futurism in a nation-state built on the dream of a socialist utopia. To this end a slogan has been pressed into service: 'mass entrepreneurship, mass innovation' (大众创业, 万众创新). The cloud of effluent that greets incoming visitors to China will gradually dissipate, autonomous cars will drive people to their destinations, and robots will care for the sick and the aging population.

2 See https://www.globalinnovationindex.org/.

3 Alec Ross, *The Industries of the Future: How the Next 10 Years of Innovation Will Transform Our Lives at Work and Home* (London: Simon & Schuster, 2016).

4 This is the so-called lucky country syndrome, a term applied to Australia in an ironic sense to draw attention to its wealth of natural resources and its reliance on this income stream. For a critique see Paul Cleary, *Too Much Luck: The Mining Boom and Australia Future* (Collingwood: Black Inc, 2011).

5 New Perspectives – What's Wrong with TED Talks? Benjamin Bratton at TEDxSanDiego 2013 – Re:Think, available at https://www.youtube.com/watch?v=Yo5cKRmJafo.

6 Techno-utopianism is probably a term we need to critique carefully; some of the most thought-provoking works on our future invoke ideas of robotics and shifts in modes of employment. See for instance Ross, *The Inducsties of the Future*.

7 Tung-hui Hu, *A Prehistory of the Cloud* (Cambridge, MA: MIT Press, 2015).

If 'innovation is the soul of a nation's progress' (创新是一个民族进步灵魂), as a large neon sign in Shanghai's East Nanjing Road tells us, and China *must* innovate, what about creativity? Creativity is not the same thing as innovation and the Chinese government knows this, placing the development emphasis on the latter; that is, the government does not call for 'mass creativity' (大众创意). Creativity, defined in the Western tradition, and celebrated in many creative pursuits, feeds off dissonance, disorder, and decadence. I will argue that the forms that creativity takes in China today are conditioned by social norms and language conventions, some dating back thousands of years, and others that have come with globalisation, bringing Chinese creative practitioners into a closer dialogue with international communities – for instance, art biennials, independent film festivals. Accordingly, it is important to bear in mind that historical perspectives do not determine people's behaviour. It is not my intention therefore to mount a case for Chinese exceptionalism, to propose that Chinese creativity is in some way unique. I will argue, however, that an examination of traditional Chinese culture can provide us with a way to move away from a Western focus on creation ex nihilo, the view that creativity comes from some external source, for instance a creator's ego, and is manifest in exceptional or 'gifted' individuals. In exploring the 'affordances' of creativity I will argue that creativity 'cannot be thought of outside the relationship between a person and the world, both social and material'.[8] The extension of this argument is that creativity is embedded in nature and is inherently 'participatory'.

I begin the chapter with a brief discussion of language. Carsten Herrmann-Pillath observes that 'Language has the emergent property of *meaning*, which in turn becomes a *causal* force in the world'.[9] In traditional China, control of language facilitated the generation of rituals and patterns that informed conventional models of social knowledge. In this way, cultural norms worked to limit creative expression. In the following sections, I will argue that specific forms of instrumental knowledge, introduced from outside of China and filtered through official channels, have become central to China's innovative nation strategy: these forms of knowledge come with historical perspectives, for instance liberal pluralism and, more recently, neo-liberalism. I illustrate this process with an account of my chance encounter

8 Vlad Glăveanu, 'Things', in *Creativity – A New Vocabulary*, ed. V. L. Glăveanu, L. Tanggaard and C. Wegener (London: Palgrave, 2016a), 169.

9 Carsten Herrmann-Pillath, *The Economics of Identity and Creativity: A Cultural Science Approach* (St Lucia: University of Queensland Press, 2010), 22. Emphasis in original.

with China's creative *zeitgeist* in the early 2000s. In the following section, I look at some of the cognitive dissonances that occur when knowledge society discourses found in pluralist societies are introduced into China, noting the British consultant John Howkins' best practice advice to China. Finally, I turn to the theme of 'affordance', a way to understand how situated participatory creativity functions in China as a never-ending process and how it is evident in the *shanzhai* movement.

Language, Names, and Knowing

Dissonance is part of Chinese culture and society and has been for more than two thousand years. While recent manifestations of online activism by 'netizens' have drawn a great deal of scholarly attention,[10] examples of dissonant behaviour abounded in the pre-internet era – the big character posters (大字报) of the anti-Rightist Movement in 1957, the ideological fervour of the Great Proletarian Cultural Revolution (1966-1976), the aspirations of the Democracy Wall Period (1978-1981), and the political rancour surrounding events in Tian'anmen Square in May-June 1989.[11] Throughout Chinese history, while dynasties rose and fell, scholars and intellectuals played key supporting roles. In ancient China during the so-called Warring States period (475-221 BCE) many schools of thought existed, metaphorically a hundred contenders, according to the well-known maxim.[12] The main schools congregated around the Confucians (Confucius 孔子, Mencius 孟子, and Xunzi 荀子), the Daoists (Laozi 老子and Zhuangzi 庄子), the Mohists (Mozi 墨子) and the Legalists (Guan Zhong 管仲 and Shang Yang 商鞅). Another 'school' of some notoriety was the School of Names (*ming jia* 名家), arguably the most creative, although that accolade is shared with the Daoists, and particularly with Zhuangzi, whose work was littered with creative associations. Laozi, the father figure of Daoism and author of the *Dao Dejing* (道德经), likewise was inclined towards allusions. Imbued with naturalism, the idea that natural laws govern behaviour, Daoism offered a far less authoritarian worldview than the culturally informed schools of Confucianism.

10 See Guobin Yang, ed., *China's Contested Internet* (Copenhagen: NIAS Press, 2015).

11 Shao Jiang, *Citizen Publications in China before the Internet* (New York: Palgrave Macmillan, 2015).

12 The maxim A Hundred Schools of Thought Contending *baijia* zhengming百家争鸣 referred to the various schools of philosophy that sprang up and flourished during the Spring and Autumn Warring States Period.

What was common to all schools of thought in China was the concept of the *dao* (道), broadly understood as the 'way' or 'path'. Another much used term was *zhi* (知), referring to the acquisition of knowledge or information. The two words combined constitute *zhidao* (知道) 'to know', literally 'know the way.' The key questions were: whose way and which way? Historically, the contending in Chinese thought depended to a large extent on defining the *dao*, and in the process eliminating alternative *dao*; hence the presence of multiple schools, and the need to provide advice to rulers, a process that continues to this day. The Confucians believed that correct names were essential. When Confucius' disciple asked the master, what is the first thing to do in starting a government, Confucius replied 'rectifying names'.[13] The Daoists on the other hand argued that language was always inadequate to describe the *dao*. Confucianism with its emphasis on ritual (*li* 礼) and governance eventually became the key element of Chinese culture, despite its repudiation by the Communists during the Cultural Revolution (1966-1976). Confucius himself put great emphasis on benevolence (*ren* 仁), which combined with ritual, would lead to self-cultivation of the superior man (*junzi* 君子) and social harmony. The practice of the rites, according to Michael Puett (2001), 'involves the establishment and maintenance of a whole range of perceptual norms concerning movement, behaviour, expression, language, dress and colour' (p. 13).[14] Creativity as such is weighted towards aesthetic rather than disruptive forms.

An important element of Confucius' teaching is the *junzi* ideal, which is mentioned over 100 times in *The Analects*. Charlene Tan has elaborated on how this concept pertained to creativity in China during Confucius' time, and later throughout East Asia.[15] Everyone could, in theory at least, become a *junzi* through moral self-cultivation, although it should be noted that in Confucius' time, and throughout most of imperial China, this realisation referred specifically to males. Tan suggests that *the junzi* embodied creativity in three ways: first, by interpreting experiences and constructing personal meanings within their sociocultural context.[16] The *junz*i might utilise 'his' imagination to respond in an 'appropriate' manner to a range of situations. Second, the *junzi* had a moral role to broaden the *dao*, thereby bringing

13 *Lunyu*, Analects 13.3.

14 Michael J Puett, 2001. *The Ambivalence of Creation: Debates Concerning Innovation and Artifice in Early China* (Stanford, Cal: Stanford University Press, 2001), 13.

15 Charlene Tan, 'Understanding Creativity in East Asia: Insights from Confucius' Concept of *Junzi*', *International Journal of Design Creativity and Innovation* 4.1 (2016): 51-61.

16 Ibid., 55.

about social harmony and benevolence (*ren*). The third point is that the *junzi* 'manifests creativity by bringing about an intermediate, rather than a radical domain change in society'.[17] In other words, the kind of creative interventions made were not likely to break with tradition. In regard to influence Weihua Niu notes that 'A creative person has to be an exemplary individual, the Confucian ideal, a person with character, to have an impact on the environment'.[18] Here, we see the close link between culture and creativity, and the idea that the latter is somehow exceptional, situated apart from the natural world.

Skills, Propaganda, and Know-to

Tan's account of Confucian-style creativity arguably dichotomises East and West: Confucius versus Plato, or even Kant. Tan is unapologetic in her view that Eastern-style creativity is more appropriate to the modern world in that it 'preserves social harmony'.[19] Furthermore, she believes that the emphasis on inculcating skills 'in East Asian societies' is in stark contrast to Western education, which values 'exploration and experimentation'.[20] While we need to be wary of exceptionalism, it is necessary to account for the determining role of culture, and the persistence of cultural norms. In *Created in Japan*, a book written at the moment when Japanese electronic industries were pushing into international markets, Sheridan Tatsuno coined the term 'creative mandala' to describe a process from '(Idea) Search' to 'Nurturing' to 'Breakthrough' to 'Refinement' and 'Recycling'.[21] In this view, creativity is an unending process where the value chains extends to further product applications, leading Tatsuno to conclude that: 'Whereas the Japanese recognise their weaknesses – basic research (ideas exploration) and breakthrough thinking (idea generation) – and are trying to connect them, Westerners mistakenly believe that they have a monopoly on creativity [...] if this myopia continues, the Japanese could eventually master the entire

17 Ibid., 56.

18 Weihua Niu, 'Confucian Ideology and Creativity', *The Journal of Creative Behaviour* 46.4 (2013): 282.

19 Tan, 'Understanding Creativity in East Asia', 59.

20 Ibid., 58.

21 Such an iterative process is found in the television industry. TV formats follow such a process of refinement and recycling. Sheridan Tatsuno, *Created in Japan: From Imitators to World-Class Innovators* (New York: Harper Business, 1990).

mandala of creativity, which would have a devastating impact on Western industry in the 21st century'.[22]

Without doubt, the impacts on Western industry in the 21st century are now coming from mainland China. Business analysts George Yip and Bruce McKern say that Chinese companies have honed their technological capabilities through incremental innovation, having acquiring state of the art skills in the school of cost reduction, and through adapting and improving.[23] Charlene Tan's distinction between 'evolutionary' (East Asian) and revolutionary (Western) creativity is evident in how products are fine tuned to the needs of consumers. While China's record of innovation is conspicuous by an absence of 'frontier innovations',[24] it is probably only a matter of time before this kind of innovation happens. But what about the influence of China on the West? Drawing on Howard Gardner's accounts of Chinese education Tan conjectures that Westerners can learn much from the skills-focused approach that allows young Chinese 'the freedom to create powerful new messages which can be understood by others'.[25] The implication here is that Chinese-style 'moral-social' and 'incremental dimension creativity' is more appropriate to our modern globalised and multicultural world.

A question arises from this discussion of normative behaviour. Who determines the contemporary *dao*, the guidelines of conduct? Is the 'creative' *junzi* the archetype of today's apparatchik, the ubiquitous propaganda official? In many respects, Propaganda Department[26] officials are invested with a special power: they can greenlight projects that may be problematic in some way. They take the ruling *dao*, for instance, the extant Five-Year Plan and adjudicate on the appropriateness of the project. So rather than seeing such officials as arch conservatives, they are in essence 'can-do' people, boundary riders who make things happen. They are important in urban development projects,[27] which often marry investors with state-sanctioned concepts – and in think-tanks that seek to promote ideas within the parameters of ideology. In recognising the role of such persons, we notice

22 Tatsuno, *Created in Japan: From Imitators to World-Class Innovators*, 54.

23 George Yip and Bruce McKern, *China's Next Strategic Advantage: from Imitation to Innovation* (Cambridge, MA: MIT Press, 2016).

24 See John Schmid and Fei-ling Wang, 'Beyond National Innovation Systems: Incentives and China's Innovative Performance', *Journal of Contemporary China* 26.104 (2017): 280-296.

25 Howard Gardner, 1989, cited in Tan, 'Understanding Creativity in East Asia', 58.

26 Sometimes called Publicity Department.

27 See Michael Keane, *China's New Creative Clusters: Governance, Human Capital and Investment* (London: Routledge, 2011).

a tension between 'what China needs' and 'what China wants'. In short, China *needs* the kind of creative experimentation that has become the norm in pluralist societies. Many scholars, policy makers, businesspersons and artists recognize this need and are willing to express it to foreigners. What the Chinese government *wants*, however, is a 'harmonious society'.

Another way of understanding this dimension of Chinese creativity is through distinctions that pertain *between* kinds of knowing; whereas 'knowing-that' (or prescriptive knowledge)[28] and 'knowing-how' (tacit knowledge) are generally assumed to be dominant modalities of knowledge, we can identify a third variant, 'knowing-to', which comes into play at important times; a person, for instance, a *junzi* in traditional China – or a propaganda official today – might wish to promote an idea but would require an appropriate time and situation to do so. Creativity pertains to choosing the moment to act, to offer advice for instance. Knowing-to manifests in four circumstances: first, anticipating outcomes (understanding the effects of an action or a policy); second, timeliness (making one's move at the right time); third, context (working in a way that takes account of other's political obligations); and fourth, understanding weightiness (knowing the relative weight of regulations).[29]

Creativity: In Situ

The argument so far can be summarised as follows. Cultural tradition has largely informed extant forms of creative expression in China, past and present. Now, with opportunities for collaborative production combined with knowledge sharing, for instance, *shanzhai* and maker culture, cultural formats are multiplying. Yet, at the same time, creativity is still widely viewed as the province of exemplary individuals generating exceptional events, ideas and products. Such a view is very similar to that propounded in the Western tradition dating back to Plato.[30] Yet, this view does not square with how creativity, and the creative industries, have evolved in China over the past two decades. The ethos of 'creating' in China has a distinctive 'local' character, namely, the informal copying practices known collectively as *shanzhai*. As I discuss later in the chapter, *shanzhai* culture

28 See Gilbert Ryle, *The Concept of Mind* (Chicago, IL: University of Chicago Press, 1949).

29 Stephen Hetherington and Karyn L. Lai, 'Knowing-how and Knowing-to', in *The Philosophical Challenge from China*, ed. Bryan Bruya (Cambridge, MA: MIT Press 2015), 279-301.

30 See Theodore Gracyk, *The Philosophy of Art* (London: Polity, 2012), 44.

is often celebrated by grass-roots communities.[31] In effect, it reconciles knowledge sharing with local ingenuity, emphasising frugality and utility. The *shanzhai* phenomenon shows that creativity is collaborative. Roger Ames and David Hall propose that creativity is constituted of '*self*-creativity and *co*-creativity'.[32] As Will Buckingham notes, citing Stuart McLean[33]: 'this permits us to rethink creativity outside of the exclusive preoccupation with cultural creativity'.[34] Buckingham believes that it is participation in broader processes, beyond narrowly circumscribed 'cultural domains', that constitutes creativity.[35] In short, creativity is *in situ* – and it is fundamentally relational.[36]

How, then, do I situate myself? And what is my relationship? My involvement in China's creative *zeitgeist* is purely accidental, serendipitous. In 2003, when I began my research on this topic,[37] like many others I believed that the Chinese government was unwavering in its control of creative expression. Yet, there were signs that the nation was 'liberalising' in response to the WTO accession. In August 2003, a small forum took place at the Queensland University of Technology (QUT) to investigate the internationalising of the creative industries, as they were then playing out. During this forum, Jing Wang from MIT expressed scepticism as to whether the creative industries could take off in China considering the state's investment in the cultural industries.[38] In other words, culture and ideology would prevail over creativity. As it turned out there would be many twists and turns.

Later that year, I travelled to Beijing to pursue some contacts for our research. One of the first persons I sought out was a scholar I knew from

31 See Fan Yang, *Faked in China: Nation Branding, Counterfeit Culture and Globalization* (Bloomington, IN: Indiana University Press, 2015).

32 Roger Ames and David Hall, *Dao De Jing: Making This Life Significant – A Philosophical Translation* (New York. Ballantine, 2004), 17. Emphasis in original.

33 Stuart McLean, 'Stories and Cosmogonies: Imagining Creativity Beyond "Nature" and "Culture"', *Cultural Anthropology* 24.2 (2009): 213-45.

34 Will Buckingham, 'Participation, Pattern and Change: Creativity in Liu Xie and Lucretius', *NTU Studies in Language and Literature* 29 (2013): 7.

35 Ibid., 21.

36 See Ames, 'Collaterality in Early Chinese Cosmology: An Argument for Confucian Har mony (He 和) as Creatio in Situ', *Early China* 37 (2014): 445-70; Silja Graupe, 'Do Daoist Principles Justify "Laissez Faire Policies" A Critical Examination of "Market Daoism"', *International Journal for Field-Being* 6.1 (2007): 1-16; Niu, 'Confucian Ideology and Creativity'.

37 The research project was called *Internationalising the Creative Industries: China the WTO and the Knowledge-based Economy.* Aside from myself, the team included John Hartley, Stephanie Hemelryk Donald, Stuart Cunningham, Terry Flew, and Christina Spurgeon.

38 Jing Wang, 'The Global Reach of a New Discourse: How Far Can "Creative Industries" Travel?', *International Journal of Cultural Studies* 7.1 (2004): 9-19.

the Chinese Academy of Social Sciences (CASS) who was working on the internet in China. He introduced me to Zhang Xiaoming, the editor of a new annual publication called *The Blue Book of China's Cultural Industries*. I had cited Prof. Zhang's work in my PhD dissertation so this was a fortuitous meeting. We would work together on several projects and events over the next decade and the *Blue Book of China's Cultural Industries* would become the bible of cultural reform. Zhang had an institutional investment in the cultural industries but was interested in the creative industries insofar as the latter was being picked up in Hong Kong and Singapore. A team led by Desmond Hui from the University of Hong Kong's Cultural Policy Centre had just published *The Baseline Study of Hong Kong's Creative Industries*. Zhang introduced me to Jin Yuanpu, a scholar who had traversed from literature and cultural studies to become interested in the cultural industries and in creativity. Jin had a new centre at Renmin University called the Humanistic Olympics Research Centre.

In April 2004, I received an alert to a new opinion piece published on a new online forum devoted to the cultural industries, entitled 'Bring about a Creative Century: Take Action to Develop a Creative China'. The author, then a section head in the Internet Division of the Ministry of Culture, had produced a manifesto calling for China to adopt the UK model of creative industries, citing the problems of rote learning and lack of original thinking that were holding the nation back. Reading this paper, one line jumped off the page: 'From Made in China to Created in China' (从中国制作到中国制造).[39] I took this as a sign that there was a path forward. At the time the author, Liu Shifa, proposed three points of intersection: 'digital China', 'creative China', and 'cultural China'. Noting that one of the core problems facing China was the 'duck style feeding' (*tianya shi*) of students in educational institutions, Liu argued that China needed to make a transformation from an economy that over-emphasised learning from others – one that 'inherits tradition, follows others, copies, and brings in' other cultures – to a 'creative economy' where creativity is the priority strategy and originality is acknowledged and valued.[40] Creativity had to be instilled, he said, not drilled.

39 The article was available online for a short while before being removed. It was also published in the proceedings of a conference, The Second International Forum on China's Cultural Industries, held in Taiyuan, Shanxi in May 2004.

40 Shifa Liu, 'Implementing the Creative Century Plan and Promoting the Creative China Campaign', in *Proceedings of The Second International Forum on China Cultural Industries Anthology* (Beijing: People's Daily Press, 2004), 90.

At a conference in early May that year at Renmin University organised by Jin Yuanpu many speakers speculated how the organisers in Beijing might creatively present the Olympics Opening Ceremony to the world. This was a chance to change international perceptions of China. In my presentation, I read out the words 'From Made in China to Created in China', drawing attention to this somewhat radical idea. In the audience that day was Su Tong, the director of a small consulting think-tank called the Creative China Industrial Alliance (CCIA). Su too recognised the power of this idea and we became good friends. The next day, he introduced me to the author of the manifesto who was at the time affiliated with the CCIA. I soon found out that Su is an autodidact, a person that picks up ideas and repackages them, finding associations that only a lateral thinker could. His predilection for word associations in particular were sometimes lost on cultural officials who only responded to the government slogans of the day. Over the following year he developed an idea called The China Code, an attempt to bring traditional Chinese ideas into contemporary place branding. I was grateful when he provided me with the cover illustration of my first book *Created in China: The Great New Leap Forward*.

Over the next several years, the cultural industries assumed greater national significance. More Blue Books proliferated, containing data and advice.[41] Creativity was marginalised and many scholars were advised to avoid talking about 'creative industries'. A number were rewarded with cultural industry research grants and funding for cultural industry research centres. Despite this pressure to conform, creativity found its way into the lexicon of business, into mainstream media, and occasionally into education debates. Li Wuwei, who was then vice-chair of the Revolutionary Guomindang, China's 'official opposition party' was forthright in declaring that 'creativity was changing China',[42] a position that often saw him in conflict with Ministry of Culture officials. Interestingly, Li Wuwei's given name (*wuwei*) literally means 'have no fear'. It is also a homophone for 'effortless action', the Daoist term *wuwei* that was translated into economic jargon as 'laissez-faire'.[43] Although now officially retired, Li's advocacy of creativity, diversity, and tolerance stands him apart from China's Confucian-Leninist cadres.

41 Zitong Qiu, 'The Remedy Paradigm: Blue Books and the Administration of cultural industries in China', in *The Handbook of Cultural Industries and Creative Industries in China,* ed. Michael Keane (Cheltenham: Edward Elgar, 2016), 15-26.

42 Wuwei Li, *How Creativity is Changing China*, eds. Michael Keane, trans. Marina Guo and Li Hui (London: Bloomsbury Press, 2011).

43 See Ken McCormick, 'The Tao of Laissez-Faire', *Eastern Economic Journal* 25 (1999): 331-41.

3.1.1 Book cover of *Created in China: the Great New Leap Forward* : concept and design by Su Tong

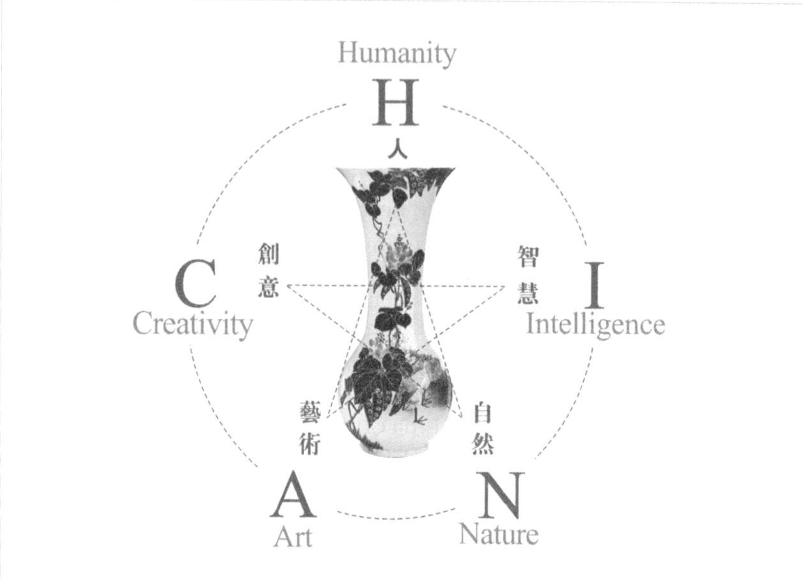

On Affordances, Best Practice, and *Creatio in Situ*

I have suggested that Chinese creativity is best understood as creation *in situ*; that is, relational – we are always situated. One way to understand this is through the notion of 'affordance'.

Glăveanu[44] traces the concept of affordance to James J. Gibson[45], a founding father of ecological psychology. Gibson's affordance is essentially concerned with an action that is relational to both the environment and humans. Perhaps a simpler way of saying this is that the environment is not 'external' to humans. In Gibson's original environmental usage the idea of an affordance is reflected in the Chinese expression *tianren heyi* (天人合一), which can be rendered as 'the continuity between the religious, natural, and cultural context, and the human experience'.[46] The ancients believed a co-creative potential exists in the world around us, the world that we inhabit; that is, we

44 Glăveanu, 'Affordance', in *Creativity – A New Vocabulary,* ed. V. L. Glăveanu, L. Tanggaard and C. Wegener (London: Palgrave, 2016b), 10-17.

45 James Gibson, 'The Theory of Affordances', in *Perceiving, Acting and Knowing,* eds. R.E. Shaw and J. Bransford (Hillsdale, NJ: Lawrence Erlbaum Associates, 1977), 67-82.

46 Ames, *Confucian Role Ethics. A Vocabulary* (Hong Kong: The Chinese University Press, 2011), 53.

are not separate individuals nor do we have a specific identity. We are more than individualised actors: for instance, I might be a son, a brother, a professor, a human being, a cosmic entity.[47] We each have a unique individuality but we also have an unbroken continuity with our significant others, and perhaps, in the internet age, of friending less significant others.

In the same way, things (objects, artefacts) are not fixed in time and space. When applied to objects, such as devices, an affordance is neither dependent on the physical property of the object nor is it completely relative to the intentions of the human actor, or user. Consider the famous divergent thinking test 'how many uses are there for a brick?' – e.g. hammer, doorstop, bookend, weapon, or artwork. This reveals the contextual nature of things, objects, and artefacts – and, by extension, creativity. The contemporary understanding of affordances, namely the digital capabilities on our mobile phones, illustrates how technology affords us opportunities to become more adaptive, and arguably more creative. Glăveanu coins an alternative definition of creativity, not about generating ideas but: 'expanding our action possibilities by perceiving or creating new affordances and exploiting affordances in new ways'.[48]

In the same vein, the emergence of the discourse of creativity in China affords people more ways of acting. What kind of creative advice should one provide for China? Over the past two decades, many foreign consultants have come forward to offer advice, often providing cameos at cultural industry expos. John Howkins is a British consultant best known in China as the 'father of the creative industries'. He has presented at many high-level conferences as well as provided advice to regional development projects. What Howkins does best is illustrate how creativity works in liberal democracies.

Howkins introduced his idea of the creative economy in 2001. More recently, he has described a 'creative ecology'.[49] His creative ecology operates on three principles: 'everyone is creative'; 'creativity needs freedom', and 'freedom needs markets'. This makes sense in London, Sydney, and Tokyo. First, the idea that 'everyone' is creative is not accepted in China. The Chinese speak of talented people (*rencai* 人才) on the one hand and 'embodied human capital' (*suzhi* 素质) on the other. The former need to be identified and rewarded. The term 'low *suzhi*' refers to people who lack the necessary social and cultural aptitudes to elevate themselves. In contrast

47 See Ames, 'Collaterality in Early Chinese Cosmology', 459.

48 Glăveanu, 'Affordance', 16.

49 John Howkins, *Creative Ecologies: Where Thinking is a Proper Job* (Brisbane: University of Queensland Press, 2009).

to talented people, government identifies such low *suzhi* groups as needing more directed education and supervision. Prejudices run deep. Andrew Kipnis illustrates this with an example of an educator who associated low *suzhi* with a region, legitimising the idea that students from certain places are backward despite their test scores.[50] One might argue that such prejudgement is not egalitarian. Then again, egalitarianism is not a concept associated with Confucianism although Confucius himself said that teachers should not discriminate in favour of those from upper echelons of society. Nevertheless, self-cultivation via ritual had its rewards. As mentioned earlier, the creative person in traditional China should be an 'exemplary individual', preferable well situated so that they can contribute.

In a broad sense, learning is something that all can acquire. Mencius, Confucius's most ardent supporter, proclaimed that all people are born with the potential to be good. By 'good' Mencius meant 'becoming' a realised person, which takes time. It's hardly surprising that young people are deferential to elders. Marina Zhang contends that the 'cognitive patterns' of Chinese people are dominated by 'customary thinking', which she says 'manifests in decision making and action'.[51] She illustrates this cultural dimension in four aspects: pursuit of equality; that is, it is hard for an individual to take the initiative to make a change; respect of the status quo; pursuit of 'feeling' rather than fact; and obeying higher powers or personal relationships instead of rules. Indeed, creative ideas are more likely to be recognised in China if they come from a person with some social or professional standing. It's a case of 'knowing-to'.

Second, the truism 'creativity needs freedom' has a distinctive liberal ethos. Howkins writes: 'It (a creative ecology) needs freedom from constraints such as physical want, hunger, prejudice, censorship and unhelpful education systems'.[52] China misses out here. Of these freedoms, extreme state censorship and the structure of the education system conspire against people challenging 'conventional wisdom'. Of course, there is a flip side to this argument; that is, that authoritarianism engenders creative resistance, a practice that journalists call 'edge-ball' (*ca bian qiu*). The Cultural Revolution for instance was a time of great political upheaval that gave birth to unprecedented creative expression. Today, the internet in China is over-populated by parody, much of it generated by people under forty.

50 Andrew Kipnis, 'Neoliberalism Reified: *Suzhi* Discourse and Tropes of Neoliberalism in the People's Republic of China', *Journal of the Royal Anthropological Institute* (*N.S.*) 13 (2007): 383-400.

51 Marina Zhang, *China 2.0* (Singapore: John Wiley and Sons, 2010), 40.

52 Howkins, *Creative Ecologies*, 130.

Yet, this variant of creative expression is extremely fragile, and in the age of surveillance 'perpetrators' are tagged in data banks as potential troublemakers.

Third, 'freedom needs markets' conjures up the vision of a typically liberal democratic polity. Many within the academic community might conjecture that this statement celebrates neo-liberalism. However, Howkins does not argue for any privileged version of governance. Rather the proposition here is that creative work needs markets, and, in particular, social marketplaces where people are free 'to exchange ideas, to enquire, to check the truth of what is happening, to explore our own paths'.[53] China has abundant social networks; yet the truth value of these networks remains clouded due to the operational failure of the previous principle, 'creativity needs freedom'.

Understanding the Affordances of Creativity in China

Introducing development concepts that are endowed with the characteristics of pluralism into China is a process fraught with uncertainty. Moreover, if these principles don't work in China, or at least come up against political and cultural barriers, how then can we make sense of Chinese creativity, for instance the much-celebrated practice of *shanzhai* manufacturing discussed elsewhere in this publication? Is there an identifiable variant of Chinese creativity? Another way of understanding these questions is to ask: What affordances might the discourse of creativity produce in China?

While technological innovation has become the default discourse of the Chinese Thirteenth Five Year Plan, creativity finds expression in manifold ways. Its resilience is reflected in *shanzhai*, sometimes referred to negatively as 'copycatting'. In a similar way that the state attempted to disempower the discourse of creativity that was finding its way into business and local government in the first decade of the millennium, *shanzhai* was deemed to be the workings of 'renegades on the frontiers of innovation' who were undermining the image of China, and Chinese people, as innovative.[54] Fan Yang says that whereas the state actively constructs the national brand image for the people, now presumably an 'innovation-oriented nation', *shanzhai*, being a grass-roots phenomenon, is actually more representative of the

53 Ibid., 134.

54 Michael Keane and Elaine Jing Zhao, 'Renegades on the Frontier of innovation: the *Shanzhai* Grassroots Communities of Shenzhen', *Eurasian Geography and Eco nomics* (Special issue on Creative Economy) 53.2 (2013): 316-30.

'people'.[55] The *shanzhai* movement illustrates creation *in situ*, an ongoing, continual process of making, remaking, and recirculating versions. *Shanzhai,* like our own journeys in life, our own personal *dao* (道), is never complete.

While innovation is the default setting as China gears up for the digital challenge, 'creativity' remains useful to the government: it has great economic and 'soft power' dividends. It is certainly hard for an individual to express radical disruptive creativity in China. Risk taking is not tolerated and the envelope cannot be pushed too far. It's more a matter of 'knowing-to': when, where, and how to express ideas. But creativity certainly aligns with China's expanding sense of 'cultural power' (*wenhua qiangguo* 文化强国). Linked with economic and social development, it *affords* greater space for expression; it *affords* shanzhai businesses a way to circumvent intellectual property laws while being inventive; and it *affords* a mechanism for online users to share contentious ideas.

As Li Wuwei has reminded the world, creativity *is* changing China, but it is an ongoing process; the change began with cultural reforms in the 1980s, rapid commercialisation of media occurred in the following decade, and theme parks and cultural quarters appeared in large numbers in the early 2000s.[56] Now the buzz is incubators, maker spaces and digital hubs. They bear grandiose names like Dream Town, Cloud Valley, and Sensor Valley. The Internet of Things is already the next big thing. But along with the digital dreaming we see the encroachment of surveillance technology, a cloud of servers and networks that effectively track and monitor the expression of individuals and communities, ironically in the name of preserving a 'harmonious society'. In a world where cultural misunderstandings and trade wars are making news headlines, the divergences between East and West that have diminished thanks to communication technologies are in danger of imploding.

Bibliography

Ames, Roger. 'Collaterality in Early Chinese Cosmology: An Argument for Confucian Harmony (He 和) as Creatio in Situ', *Early China* 37 (2014): 445-70.

—. *Confucian Role Ethics. A Vocabulary* (Hong Kong: The Chinese University Press, 2011).

Ames, Roger, and David Hall. *Dao De Jing: Making This Life Significant – A Philosophical Translation* (New York: Ballantine, 2004).

55 Yang, *Faked in China.*

56 Li Wuwei, *How Creativity is Changing China.*

Buckingham, Will. 'Participation, Pattern and Change: Creativity in Liu Xie and Lucretius', *NTU Studies in Language and Literature* 29 (2013): 1-23.

Cleary, Paul. *Too Much Luck: The Mining Boom and Australia Future* (Collingwood: Black Inc, 2011).

Gardner, Howard. *To Open Minds: Chinese Clues to the Dilemmas of Contemporary Education* (New York: Basic Books, 1989).

Gibson, James. 'The Theory of Affordances'. In *Perceiving, Acting and Knowing*, eds. R.E. Shaw and J. Bransford (Hillsdale, NJ: Lawrence Erlbaum Associates, 1977), 67-82.

Glăveanu, Vlad. 'Things'. In *Creativity – A New Vocabulary*, ed. V. L. Glăveanu, L. Tanggaard and C. Wegener (London: Palgrave, 2016a), 164-71.

—. 'Affordance'. In *Creativity – A New Vocabulary*, ed. V. L. Glăveanu, L. Tanggaard and C. Wegener (London: Palgrave, 2016b), 10-7.

Gracyk, Theodore. *The Philosophy of Art* (London: Polity, 2012).

Graupe, Silja. 'Do Daoist Principles Justify "Laissez Faire Policies" A Critical Examination of "Market Daoism"', *International Journal for Field-Being* 6.1 (2007): 1-16.

Herrmann-Pillath, Carsten. *The Economics of Identity and Creativity: A Cultural Science Approach* (St Lucia: University of Queensland Press, 2010).

Hetherington, Stephen, and Karyn L. Lai 'Knowing-how and Knowing-to'. In *The Philosophical Challenge from China*, ed. Bryan Bruya (Cambridge, MA: MIT Press 2015), 279-301.

Howkins, John. *Creative Ecologies: Where Thinking is a Proper Job* (Brisbane: University of Queensland Press, 2009).

—. *The Creative Economy: How People Make Money from Ideas* (London: The Penguin Group, 2001).

Hu, Tung-hui. *A Prehistory of the Cloud* (Cambridge, MA: MIT Press, 2015).

Jiang, Shao. *Citizen Publications in China Before the Internet* (New York: Palgrave Macmillan, 2015).

Keane, Michael. *China's New Creative Clusters: Governance, Human Capital and Investment* (London: Routledge, 2011).

—. *Created in China: The Great New Leap Forward* (London: Routledge, 2007).

Keane, Michael, and Elaine Jing Zhao. 'Renegades on the Frontier of innovation: the *Shanzhai* Grassroots Communities of Shenzhen', *Eurasian Geography and Economics* (Special issue on Creative Economy) 53.2 (2013): 316-30.

Kipnis, Andrew. 'Neoliberalism Reified: *Suzhi* Discourse and Tropes of Neoliberalism in the People's Republic of China', *Journal of the Royal Anthropological Institute (N.S.)* 13 (2007): 383-400.

Li, Wuwei. *How Creativity is Changing China*. Edited by Michael Keane and translated by Marina Guo and Li Hui (London: Bloomsbury Press, 2011).

Liu, Shifa. 'Implementing the Creative Century Plan and Promoting the Creative China Campaign'. In *Proceedings of The Second International Forum on China Cultural Industries Anthology* (Beijing: People's Daily Press, 2004), 87-93.

Lundvall, Bengt-Åke, Björn Johnson, Esben Sloth Andersen, and Bent Dalum. 'National systems of production, innovation and competence building.' *Research policy* 31.2 (2002): 213-231.

McCormick, Ken. 'The Tao of Laissez-Faire', *Eastern Economic Journal* 25 (1999): 331-41.

McLean, Stuart. 'Stories and Cosmogonies: Imagining Creativity Beyond "Nature" and "Culture"', *Cultural Anthropology* 24.2 (2009): 213-45.

Mitchell, William, Alan Inouye, and Marjory Blumenthal. *Beyond Productivity: Information Technology, Innovation and Creativity* (Washington, DC: The National Academies Press, 2003).

Niu, Weihua. 'Confucian Ideology and Creativity', *The Journal of Creative Behaviour* 46.4 (2013): 274-84.

Puett, Michael J. *The Ambivalence of Creation: Debates Concerning Innovation and Artifice in Early China* (Stanford, Cal: Stanford University Press, 2001).

Qiu, Zitong. 'The Remedy Paradigm: Blue Books and the Administration of cultural industries in China'. In *The Handbook of Cultural Industries and Creative Industries in China,* ed. Michael Keane (Cheltenham: Edward Elgar, 2016), 15-26.

Ross, Alec. *The Industries of the Future: How the Next 10 Years of Innovation Will Transform Our Lives at Work and Home* (London: Simon & Schuster, 2016).

Ryle, Gilbert. *The Concept of Mind* (Chicago, IL: University of Chicago Press, 1949).

Schmid, John, and Fei-ling Wang. 'Beyond National Innovation Systems: Incentives and China's innovative Performance', *Journal of Contemporary China* 26.104 (2017): 280-296.

Tan, Charlene. 'Understanding Creativity in East Asia: Insights from Confucius' Concept of *Junzi*', *International Journal of Design Creativity and Innovation* 4.1 (2016): 51-61.

Tatsuno, Sheridan. *Created in Japan: From Imitators to World-Class Innovators* (New York: Harper Business, 1990).

Wang, Jing. 'The Global Reach of a New Discourse: How Far Can "Creative Industries" Travel?', *International Journal of Cultural Studies* 7.1 (2004): 9-19.

Yang, Fan. *Faked in China: Nation Branding, Counterfeit Culture and Globalization* (Bloomington, IN: Indiana University Press, 2015).

Yang, Guobin, ed. *China's Contested Internet* (Copenhagen: NIAS Press, 2015).

Yip, George and Bruce McKern. *China's Next Strategic Advantage: from Imitation to Innovation* (Cambridge, MA: MIT Press, 2016).

Zhang, Marina. *China 2.0* (Singapore: John Wiley and Sons, 2010).

About the Author

Michael Keane's key research interests are digital transformation in China, East Asian cultural and media policy; and creative industries and cultural export strategies in China and East Asia. Michael is author or editor of seventeen books on China's media: His single authored publications are *China's Television Industry* (Palgrave, 2015), *Creative Industries in China: Art, Design and Media* (Polity, 2013), *China's New Creative Clusters: Governance, Human Capital and Regional Investment* (Routledge, 2011), and *Created in China: The Great New Leap Forward* (Routledge. 2007). Michael is director of the Digital China Lab at Curtin University. He is currently leading an Australia Research Council-funded project (DP170102176), which examines China's attempt to use 'digital power' of the internet to achieve international recognition as an 'innovative nation'.
m.keane@curtin.edu.au

3.2 Digital Payment, Vernacular Creativity, and Governmentality

Chen Siyu

Abstract

Built on the expanding e-commerce platforms and social networks, digital payment in China has facilitated the integration of online and offline economies. Apart from its business potential, digital payment also has a major impact on the cultural, social, and political landscape of contemporary Chinese society. Drawing on Jean Burgess's concept of 'Vernacular Creativity' and Michel Foucault's theorisation on governmentality, this essay explores the socio-political changes brought about by digital payment through the lens of digital red packets and Sesame Credit Scores. By probing the new forms of social interaction and governance engendered by digital payments, this essay looks at how technology intersects with social practices and political ends in shaping the everyday lives in China.

Keywords: digital payment, governmentality, vernacular creativity, digital red packets, Sesame Credit Scores

In 2016, when I went back to China for the winter holiday, I had dinner with a friend at a restaurant. After we finished, I reached into my wallet the moment my friend took out her phone. For the restaurateur, it seemed to be a pretty easy decision to make. 'Let's pay by phone', he said without hesitation, 'it saves the time looking for change'. My friend opened the WeChat App on her phone, scanned the barcode on the bill and inputted the password. No more waiting at the counter, the payment was made instantly. I got myself a WeChat Pay account the next day.

Kloet, Jeroen de, Chow Yiu Fai, and Lena Scheen (eds), *Boredom,* Shanzhai, *and Digitisation in the Time of Creative China*. Taylor & Francis Group, 2019
DOI: 10.5117/9789462984745_CH3.2

WeChat Pay, along with its closest competitor Alipay, dominate the digital payments space in China, capturing 28 per cent of all retail transaction fees.[1] These two, far-reaching applications in China enabled a total of $2.9 trillion in digital payments last year, representing a twenty-fold increase from 2012 to 2016.[2] Built on the expanding e-commerce platforms and social networks in China, the digital payments ecosystem facilitates the integration of online and offline economies, albeit increasing control and surveillance over financial flows. Driving Business to Customer (B2C) and Peer to Peer (P2P) transactions, digital payment in China supports a wide range of digital financial services, expanding financial inclusion and economic opportunity for individuals, and creating valuable new business models for companies. Apart from its business potential, digital payment also has a major impact on the cultural, social, and political landscape of contemporary Chinese society. By probing into the new forms of social interaction and governance engendered by digital payments, this essay looks into how technology intersects with social practices and political ends in shaping the everyday lives in China.

Red Packets and Vernacular Creativity

According to Huang Zheping, the Chinese internet, of which digital payment is a crucial component, has transformed from a panopticon where citizens are kept under control by a central device to a panspectron where information is increasingly being generated and shared by sensing devices that are located all around us.[3] The transformation opens up new opportunities for individuals to engage in the ordinary cultural production through the creative usage of internet. Arguing that categorisations such as 'producer', 'audience', and 'consumer' become inadequate in the context of new media, Jean Burgess defines everyday creativity practiced outside the cultural value systems of either high culture or commercial creative practice as 'vernacular

1 Better Than Cash Alliance, 'Social Networks, e-Commerce Platforms, and the Growth of Digital Payment Ecosystems in China: What It Means for Other Countries', April 2017, accessed July 20, 2017, https://btca-prod.s3.amazonaws.com/documents/283/english_attachments/Better_Than_Cash_Alliance_China_Report_April_2017_(1).pdf?1492605583.

2 Ibid.

3 Huang Zheping, 'All Chinese Citizens Now Have a Score Based on How Well We Live, and Mine Sucks', *Quartz*, February 6, 2016, accessed July 21, 2017, https://qz.com/613384/over-8-billion-red-envelopes-were-sent-over-wechat-during-chinese-new-year/.

creativity'.[4] Exploiting competencies built up through everyday experience as a user of new media, vernacular creativity articulates with the cultural shifts and media technology in contemporary China, as epitomised by the impact of the digital red packets (*dianzi hongbao* 电子红包) on social experiences.

WeChat Pay first launched the virtual red packet feature in 2014. It is an online version of the Chinese tradition of giving a red packet to celebrate the Lunar New Year. Allowing users to send each other digital red envelopes that are deposited into their digital payment account, each red packet contains up to 200 RMB, and can be posted to either individuals or groups with a customised message on the cover of the envelope. When it is sent to a group, the sender can decide whether each recipient gets an even or a randomised share of the money and specify the number of recipients. This added WeChat feature was an immediate hit. During the Chinese New Year's Eve in 2014, sixteen million digital red packets were exchanged, and the number jumped to one billion in 2015, and then eight billion in 2016. As Wu and Ma note, competition, combined with the sense of 'luck' and possible financial gains, turns money gifting into an exciting, lottery-style, game-like online social activity. In fact, this online feature has proven to be addictive.[5] At many of the dinner parties I attended during the Spring Festival of 2016, people seemed to have lost interest in talking to each other or watching the Spring Festival Gala; many of them spent a lot of time gazing intently at their mobile phones, cheering for the red packets they grabbed and sighing for the ones they missed.

While red packets were traditionally exchanged during major events, such as important festivals, birthdays, weddings, and funerals, etc., giving and receiving digital red packets has become a popular online activity in everyday life in China. In most of the WeChat groups that I am in, users frequently use a red packet to stimulate group dynamics. For instance, red packets are frequently given out to invite members to join the celebration of public holidays, group achievement, and individual success. Some of the groups even have a 'red packet night' (*hongbao kuanghuanye* 红包狂欢夜) during which those who receive the largest share of the red packet take turns to send red packets to their group members. Bringing fun and joy,

4 Jean Burgess, 'Vernacular Creativity and New Media', PhD dissertation (Brisbane City: Queensland University of Technology, 2007).

5 Ziming Wu and Xioajuan Ma, 'Money as a Social Currency to Manage Group Dynamics: Red Packet Gifting in Chinese Online Communities', in *Proceedings of the 2017 CHI Conference Extended Abstracts on Human Factors in Computing Systems* (ACM, 2017).

the gamification of exchanging red packets has become a common practice for WeChat users to sustain group connectedness. Moreover, digital red packets operate as a form of social currency among Chinese today. While showing gratitude through money is not customary, and in many cases even regarded as being not classy in the Chinese context, the entertaining nature of the digital red packets dispels such awkwardness. WeChat users attempt to induce immediate responses by sending red packets while asking for favours or inquiring about certain information in WeChat groups. In the same light, digital red packets have been widely used by Chinese to express affection. This rising phenomenon first manifested on May 20th, 2015. In Chinese, the pronunciation of 520 is similar to the phrase 'I love you'. May being the 5th month of the year, May 20th is also 5.20, making it a new Chinese Valentine's Day in recent years. On May 20, 2015, numerous users confessed their love by sending digital red packets of 5.20 RMB to their beloved ones. Many recipients of the red packets cheerfully responded by taking screenshots of the transaction records and posting them on their WeChat Moments, further fuelling the surge of the 520 red packet on that particular day. Due to a sudden and unexpected spike in payment transactions, WeChat Pay quickly had its servers upgraded. Facilitated by digital red packets, affections among Chinese are not only increasingly expressed through money gifting, but also documented and shared via social media, leading to the rising phenomenon of public display of affection (*Xiu Enai* 秀恩爱) on the Chinese internet. Constituting a 'politics of "ordinary" cultural participation',[6] digital red packets have become a field where Chinese individuals exercise vernacular creativity in their everyday lives through the articulation of media technology and popular traditions.

Sesame Credit Scores and Governmentality

Digital payment in China not only engenders new forms of social interaction through the reinvention of traditions but is also shaping the Party-state's governance strategy. As Michael Keane observes earlier in this section,[7] the rise of digitisation in China is accompanied by the encroachment of

6 Burgess, 'Hearing Ordinary Voices: Cultural Studies, Vernacular Creativity and Digital Storytelling', *Continuum* 20.2 (2006): 201-214.

7 Michael Keane, 'Creativity, Affordances, and Chinese Traditional Culture', in Kloet, Jeroen de, Chow Yiu Fai, and Lena Scheen (eds), *Boredom,* Shanzhai, *and Digitisation in the Time of Creative China* (Amsterdam, Amsterdam University Press, 2019): 253-271.

surveillance. Using the big data generated through online platforms, the Chinese government has been actively experimenting with a social credit system that include a crucial component called 'Sesame Credit Scores'(*zhima xinyongfen* 芝麻信用分). A credit-scoring service from Ant Financial, the financial wing of Alibaba which also runs Alipay, Sesame Credit Scores tracks its users' financial and consumption activities to compile individual 'social credit' scores. Whereas having a high score entitles one to small loans, a fast track visa application, and waivers on deposits, a low score can result in lower internet speeds, slower processing of bureaucratic paperwork, or access to job offers and loans denied. According to a report by zjol.com,[8] citizens are ranked based on the following five criteria:

1 Identity profile, including information and documentations about users' education and occupation, and the consumptions made under real names (booking hotels, purchasing flight tickets and insurances, etc.).
2 Behavioural preferences exhibited through online shopping, payment and bank transfers.
3 Social network, especially your impact on your circle and the number of friends with high social credit scores.
4 Credit history, including payment history, rental record and payment of mortgage, etc.
5 Capability to fulfil the contract, in which one's financial status, exhibited through one's savings, the purchase of financial products, possession of cars and houses, etc, plays a very important role in the evaluation.

Various media tips on how to boost one's credit score so as to pass numerical benchmarks for different perks. Jiemian.com,[9] for instance, advises its readers that to enhance one's credit score, apart from paying off credit card debt on time, one should also support charity work by making donations via Alipay, show one's sense of familial responsibility by paying for their online shopping, and make friends with people with high social credit scores. These media reports' stress on personal responsibility demonstrates the governmental use of obligation, peer pressure, and self-control as instruments of regulation. As Rob Aitken notes, the contemporary self-governing principles are particularly circulated through media, re-formulating the

8 Yunling Huang, 'Deciphering Sesame Credit: Why Your Sesame Credit is Lower than Others?', *zjol.com*, February 2, 2015, accessed July 21, 2017, http://biz.zjol.com.cn/system/2015/02/02/020494598.shtml.

9 Founded by China's largest newspaper group, Shanghai United Media Group, in September 2014, Jiemian is the most popular online news and social platform among Chinese middle class.

financial identities and practices of the citizens.[10] While the emphasis on obligations and repayments ensure that consumption is necessarily 'disciplined hedonism'[11] in an increasingly commercialising China, the advocacy for material support for the society and one's families responds to the state's call for constructing a harmonious China. This everyday financial governance by what Paul Langley terms as 'good financial citizens', is not limited to the payment/consumption side of everyday practices.[12] To boost their Sesame scores, people on various online forums are actively searching for high scorers and friend them on social media whereas lower scorers are consciously avoided.

Moreover, Alibaba encourages its users to share their Sesame scores and pit them against their friends on social media in return for cash prizes, which is an uncommon practice in countries with more established credit rating systems. On Baihe, China's biggest matchmaking service, fifteen per cent of its users choose to display their Sesame Credit score to attract potential partners,[13] and many use it as a criterion to weed out potential dates. By doing away with the idea that a credit score is a private matter, one's credit score in China has now become not only public, but also an important indicator of one's social status and trustability based on the level of social inclusion and exclusion exercised. Rendering moral judgements and peer pressure, the credit score system is a source of policing Chinese citizens so that they conduct themselves in line with the desire of the state. Sesame credit is part of a state-endorsed trial run for creating a national social system. These pilots are currently unconnected but are expected to be combined by 2020 under government leadership. As a financial credibility indicator as well as a compliance mechanism, the government mandatory system will incorporate financial, social, behavioural, and legal data to generate a score for every citizen. While the state council[14] claims that this social credit system will help to establish the idea of a 'sincerity culture'

10 Rob Aitken, 'The Democratic Method of Obtaining Capital – Culture, Governmentality and Ethics of Mass Investment', *Consumption, Markets and Culture* 6.4 (2003): 293-317.

11 Lendol Calder, *Financing the American Dream: A Cultural History of Consumer Credit* (Princeton, NJ: Princeton University Press, 2009), 31.

12 Paul Langley, *The Everyday Life of Global Finance: Saving and Borrowing in Anglo-America* (Oxford: Oxford University Press, 2008).

13 Clinton Nguyen, 'China Might Use Data to Create a Score for Each Citizen Based on How Trustworthy They Are', *Business Insider*, October 26, 2016, accessed July 21, 2017, https://www.businessinsider.nl/china-social-credit-score-like-black-mirror-2016-10/?international=true&r=US.

14 The State Council of the People's Republic of China, 'Planning Outline for the Construction of a Social Credit System (2014-2020)', 2016, accessed July 21, 2017, https://chinacopyrightandmedia.wordpress.com/2014/06/14/planning-outline-for-the-construction-of-a-social-credit-system-2014-2020/.

in China, many express concerns that it will affect regulatory compliance in a coercive way.[15]

Bliss or Woe?

In July 2017, on the first day of my field study in Shenzhen, I lost my cellphone. Though I quickly got a new phone the next day, one day without a phone was long enough for me to realise the extent to which my life revolves around digital payment, the risks of my accounts being hacked, and how painful it is to be deprived of access to mobile apps when, say, renting a bike, ordering a taxi, topping up my subway ticket, and paying for my meals, etc. As China's digital payment market continues to grow, our increasing reliance on it will open up new opportunities for vernacular creativity, new means of governmentality, and new hazards of security, with both potentials and challenges ahead.

This research is supported by a consolidator grant from the European Research Council (ERC-2013-CoG 616882-ChinaCreative).

Bibliography

Aitken, Rob. 'The Democratic Method of Obtaining Capital – Culture, Governmentality and Ethics of Mass Investment', *Consumption, Markets and Culture* 6.4 (2003): 293-317.

Better Than Cash Alliance. 'Social Networks, e-Commerce Platforms, and the Growth of Digital Payment Ecosystems in China: What It Means for Other Countries'. April 2017. Accessed July 20, 2017. https://btca-prod.s3.amazonaws.com/documents/283/english_attachments/Better_Than_Cash_Alliance_China_Report_April_2017_(1).pdf?1492605583.

Burgess, Jean. 'Vernacular Creativity and New Media'. PhD dissertation (Brisbane City: Queensland University of Technology, 2007).

—. 'Hearing Ordinary Voices: Cultural Studies, Vernacular Creativity and Digital Storytelling', *Continuum* 20.2 (2006): 201-214.

Calder, Lendol. *Financing the American Dream: A Cultural History of Consumer Credit* (Princeton, NJ: Princeton University Press, 2009).

15 Cheang Ming, 'FICO with Chinese Characteristics: Nice Rewards, But Punishing Penalties', *CNBC*, March 16, 2017, accessed July 21, 2017, http://www.cnbc.com/2017/03/16/china-social-credit-system-ant-financials-sesame-credit-and-others-give-scores-that-go-beyond-fico.html.

'How to Increase Your Sesame Credit Score', *Jiemian.com*, June 6, 2015. Accessed July 21, 2017. http://www.jiemian.com/article/298559.html.

Huang, Zheping. 'All Chinese Citizens Now Have a Score Based on How Well We Live, and Mine Sucks', *Quartz*, February 6, 2016. Accessed July 21, 2017. https://qz.com/613384/over-8-billion-red-envelopes-were-sent-over-wechat-during-chinese-new-year/.

Huang, Yunling. 'Deciphering Sesame Credit: Why Your Sesame Credit is Lower than Others?', *zjol.com*, February 2, 2015. Accessed July 21, 2017. http://biz.zjol.com.cn/system/2015/02/02/020494598.shtml.

Keane, Michael. 'Creativity, Afffordances, and Chinese Traditional Culture', in Kloet, Jeroen de, Chow Yiu Fai, and Lena Scheen (eds), *Boredom,* Shanzhai, *and Digitisation in the Time of Creative China* (Amsterdam, Amsterdam University Press, 2019): 253-271.

Langley, Paul. *The Everyday Life of Global Finance: Saving and Borrowing in Anglo-America* (Oxford: Oxford University Press, 2008).

Ming, Cheang. 'FICO with Chinese Characteristics: Nice Rewards, But Punishing Penalties', *CNBC*, March 16, 2017. Accessed July 21, 2017. http://www.cnbc.com/2017/03/16/china-social-credit-system-ant-financials-sesame-credit-and-others-give-scores-that-go-beyond-fico.html.

Nguyen, Clinton. 'China Might Use Data to Create a Score for Each Citizen Based on How Trustworthy They Are', *Business Insider*, October 26, 2016. Accessed July 21, 2017. https://www.businessinsider.nl/china-social-credit-score-like-black-mirror-2016-10/?international=true&r=US.

The State Council of the People's Republic of China. 'Planning Outline for the Construction of a Social Credit System (2014-2020)'. 2016. Accessed July 21, 2017. https://chinacopyrightandmedia.wordpress.com/2014/06/14/planning-outline-for-the-construction-of-a-social-credit-system-2014-2020/.

Wu, Ziming, and Xiaojuan Ma. 'Money as a Social Currency to Manage Group Dynamics: Red Packet Gifting in Chinese Online Communities'. In *Proceedings of the 2017 CHI Conference Extended Abstracts on Human Factors in Computing Systems* (ACM, 2017).

About the Author

Chen Siyu is an assistant professor at the School of Humanities and Social Sciences, Harbin Institute of Technology (Shenzhen). She conducted her field study on creative city in Shenzhen when working as a postdoc researcher for the ERC project ChinaCreative. Her research interests include urban studies, creative industries, gender and media in China.

3.3 Shopping 'Natural' and 'Local' food as Everyday Resistance

Digitisation, Platformisation, and Online Retail of Rural Products

Zeng Guohua

Abstract
This essay outlines how the fast-growing demand for trusted 'natural' and 'local' agricultural products, partly out of pervasive worries over the food safety in China, has resulted in the proliferation of online shopping as a form of social resistance. It also examines how this process has coincided with the digitalisation and platformisation of Chinese societies, especially in rural areas. Through this brief description, Guohua argues that the online retail of 'natural' and 'local' products may be seen as doubly creative: at the micro level, it is a creative self-caretaking process; at the macro level, it is a creative practice to create and maintain an apparatus of social resistance.

Keywords: everyday resistance, digitalisation, platformisation, online retail, rural products, creativity

> Last week, I received a parcel from one of my close friends. It is a gift with five kilograms of Chinese jujubes. 'Just like your Chinese navel orange (which you sent to me last year), these jujubes are "authentic", "natural" fruit from a truly "local" farm run by a relative of an acquaintance of mine. And they are not chemically processed before sending out.' He stressed 'authentic', 'natural' and 'local' on the WeChat (an all-inclusive app launched by Tencent, a leading IT company in Shenzhen, China), speaking in a city 2,000 kilometres away from me, and 1,900 kilometres from the locality where the jujubes were grown.
>
> – *An informant*

Kloet, Jeroen de, Chow Yiu Fai, and Lena Scheen (eds), *Boredom,* Shanzhai, *and Digitisation in the Time of Creative China.* Taylor & Francis Group, 2019
DOI: 10.5117/9789462984745_CH3.3

Authenticity, Unsafety, and Fakery

The pursuit of trustful 'natural' and 'local' agricultural products is growing rapidly in mainland China, partly as a result of pervasive worries about food safety in the country. The 'natural' here is more or less synonymous with 'organic', which means that the product is cultivated or raised in more 'traditional' ways, avoiding synthetic fertilisers/forages, pesticides, and in more 'natural' environments than dubious modern husbandry and aquaculture enclosures. Since the early 2000s, intensive media exposure of food safety incidents due to environmental pollution and human factors[1] have generated serious unease and distrust in the food supply, even creating a feeling of 'you can't eat anything if you keep watching the TV news (about the food safety issues)'. These worries give rise to various ways for urban individuals to creatively tackle or resist the food safety problems. Except for the common methods of paying more attention to the food source, e.g. buying food and raw materials in trusted shops and markets and choosing trusted brands or sources, there are also alternative means. These include urban gardening, weekend suburb gardening, crowdfunding organic farms, and direct buying of authentic 'local', 'natural', or 'organic' products from far away and remote areas where those categories of goods are regarded as the best.

None of these alternative means is easy, however, especially the last one. In addition to 'natural' and 'organic' features, the issue of 'locality' is, as elsewhere, also highly valued by consumers: in both online and offline markets, the products from the places believed to be home to the best products are generally more expensive than those from other sources. Given that 'fake' products (one of the most frequent forms of 'fakery' is labelling a product from a less recognised production area as an item from the most recognised localities) are rampant in many places in mainland China, consumers are likely to value the localness over the 'natural' and 'organic' features.

This demand for authenticity that is mapped onto naturalness and locality immediately raises two issues: information and logistics/accessibility. Even in the period 2010-2014, when China's online retailing volume was equal to the U.S. (it surpassed the U.S. in 2015), it was still regarded as extremely remarkable and creative when *Sina Weibo*'s (then the top social media platform in China) Big Verified Users (i.e. users registered with real name

1 Guobin Yang, 'Contesting Food Safety in the Chinese Media: Between Hegemony and Counter Hegemony', *The China Quarterly* (May 2013): 1-19.

and social identification and with a large number of followers) voluntarily helped the worried farmers to advertise their unsaleable agricultural products in, usually, remote areas on *Sina Weibo*. These promotions mostly turned the unsaleable goods into stock-outs and were deemed as exemplars of *Sina Weibo*'s, so the top Chinese social media claimed, 'egalitarian' and 'empowering' features. Quite some of these successful cases involved 'local' goods from highly recognised production localities, which means that even these most recognised products may also find it difficult to reach the customers willing to buy them. Even today, there are still occasional reports about how unsaleable agricultural goods (mostly fruits and vegetables) are left rotten in the field due to the stagnation of sales. The logistics/accessibility issue poses a similar problem, for only in the last two years has the express delivery service finally reached more than half of all townships across China, enabling people with the internet to direct buy fresh fruits, vegetables, and other agricultural products in season from the recognised production localities or from a trusted 'natural' or 'organic' farm. However, as we shall see, this process is not uncontested.

Digitisation, Platformisation, and Online Retail of Rural Products

The evolution of online shopping and buying of agricultural products, as part of the digitisation and platformisation of the retail industry and of Chinese society, may roughly be divided into three phases. Firstly, in the mid-2000s, when Alibaba's Taobao.com was launched, there were early entrepreneurs and Small and Medium Enterprises (SMEs) attempting to sell agricultural goods online. At that time, the selected items were mostly dehydrated goods and processed and packaged agricultural goods, in order to reduce the obvious shipping difficulty, since, at that time, the delivery generally demanded three to seven days. The 'genuine' local flavour and 'natural' way of cultivation of these products were commonly used as points for marketing, promotion, and advertisement.

The second phase was a transient stage roughly from the late 2000s to 2013. In this stage, the entrepreneurs and SMEs slowly broadened their product inventory to include more durable fruits and rhizomes. In addition to the shortening of delivery time, the delivery companies began to set up county-level service points and to offer special collection services during the seasons when the typical local agricultural products were ready for market. Small and alternative e-commerce websites were also established, e.g. the SuiNet established in Zhejiang province in 2010 achieved remarkable success in serving the local community by organising local retail services and providing village to village parcel delivery service.

The third phase is marked by the massive market development of leading e-commerce platforms in rural areas since 2014. In the early 2010s, Chinese online retailing was already equal to levels in the U.S. and began to reach saturation point in the cities. Meanwhile, in 2013 and 2014, the new government led by President Xi Jinping launched initiatives to accelerate the rural areas to realise the 'common prosperity' long promised by the Chinese Communist Party. Like many other initiatives,[2] these projects are aimed at generating new dynamics of economic growth, which have been slowing down after more than 30 years of growth at nine per cent and above. From 2014-2016, the government invested vast resources in improving the infrastructure of rural areas, including delivery services, IT infrastructure, big data centre construction, transportation, and the necessary funding support. Partly echoing the central government's call for 'common prosperity' and 'poverty alleviation with e-commerce', partly to further expand the market coverage to the rural areas where more than 600 million people live, JingDong, Alibaba, and Tencent respectively launched their own programmes to expand their market reach in rural areas. As a result of these massive campaigns, by 2016 they had reached almost half of the rural townships (even villages in some areas) in the countryside, thus integrating the rural areas into the e-commerce structure that had, until then, primarily flourished in the cities. While these platforms promoted industrial products to the rural area, the platforms also sold rural goods in much larger volumes than before. JingDong, as well as TMall (a Business-to-Consumer e-commerce platform run by Alibaba Group), systematically integrated the medium-sized and larger companies selling local products into their platforms and had been selling agricultural goods in large volumes, and the entrepreneurs, farmers, and small businesses continued their small-scale sales on Taobao.com, or started to explore the WeChat Mall (an online retailing module integrated in WeChat).

One of the vital changes accompanying the rapid e-commerce market coverage expansion to the township (and some villages) is the significant improvement of the delivery service: fast delivery, the adoption of low-temperature transportation facilities, dense service points in the townships and even some villages. Since 2015, the delivery companies ship fresh fruits, vegetables, poultry, fresh mushrooms, forest goods, and other rural goods

2 Hong Yu, *Networking China: The Digital Transformation of the Chinese Economy* (Champaign, IL: University of Illinois Press, 2017).

efficiently and at a relatively low rate. This improvement interplays with the proliferation of Alipay and WeChat Pay.[3] With the all-inclusive social app WeChat, through which the online sellers can advertise their products, find and maintain consumers, anyone can, theoretically, sell any rural products to most cities and towns in this vast country. By 2016, the online retail in the rural area amounted to 895 billion Chinese yuan; that is more than double the 353 billion in 2015 and accounts for 17.4 per cent of the overall online retail in mainland China and involves twenty million jobs.[4]

The evolution of online retail in the rural areas of China characterises a highly asymmetric model of digitisation: out of the 895 billion annual revenue of online retail in rural area in 2016, only 17.6 per cent (159 billion) was related to agricultural products selling from rural areas to both cities and the countryside, which is significantly lower than the industrial goods selling from the urban to the rural areas.[5] Moreover, through the above-mentioned evolution, one can also witness a process of platformisation in which both the economy and society of the rural areas have been increasingly reorganised by leading online retailing platforms. A political economy analysis may alert us to the growing dominance of consumerism, platform capitalism,[6] or the emergence of a platform society[7] in the rural areas, as well as in mainland China in general. However, when viewed from a different perspective, a platform like *WeChat* and *Taobao* can also enable everyday resistance to both the unsafety and 'fakery' of the food supply in China.

Online Shopping as Everyday Resistance

The digitisation and platformisation of the retail of agricultural products opens up new possibilities for people who are concerned over food safety to obtain the products they trust and want to buy. Many of the aforementioned alternative means for resisting food 'forgery' and 'fakery' are becoming easier,

3 See also Siyu Chen in this volume.

4 Ministry of Commerce of the People's Republic of China, 'The 2016 Report on China's Online Retail', 2016, p. 33, accessed August 24, 2016, http://images.mofcom.gov.cn/dzsws/201706/20170621110205702.pdf.

5 The 46.99%, i.e. 415.34 billion. Ibid.

6 Nick Srnicek, *Platform Capitalism* (Cambridge, UK: Polity Press, 2017).

7 Thomas Poell and José van Dijck, 'Constructing Public Space: Global Perspectives on Social Media and Popular Contestation – Introduction', *International Journal of Communication*, 10 (2016): 226-34.

especially for the direct buying of food or raw materials from recognised production localities. Small farms that cultivate in the 'traditional' and 'natural' way, medium or large companies emphasising the authenticity of 'localness', and various sized organic farms and agricultural social enterprises are all available on the retail platforms and they sell goods with quick delivery services assisted by smoothly operating delivery enterprises. The opening story of this essay shows that the accessibility of 'local' agricultural products is not difficult anymore, and that social media, especially WeChat, also provide crosschecks and evaluations of the quality and the localness of those goods.

The digitisation and platformisation also provide new and creative means for the pursuit and guarantee of 'naturalness' or 'organicness'. Most organic farms now provide very detailed, visually rich, and creative descriptions of the production process of the agricultural products on the social media and online retail platforms. Some of them even set up real-time online streaming video cameras monitoring the farm activities or the processing of the products. This increases the buyers' trust in the 'organicness' substantially. The crowdfunding for organic farms and agricultural social enterprises is diversifying, moving from predominantly offline to online engagements, from partnership and membership to flexible preorder and decentralised operation. Inevitably, buyers are very willing to post their experiences of engaging with these enterprises in WeChat and on other social media platforms, which again attracts more buyers as indicated in the opening story.

Both direct buying and 'organic' farming, as well as other practices such as urban gardening and weekend suburban gardening, provide a feeling of everyday ritual,[8] resisting the rampant unsafety and fakery of food production in mainland China. The searching for and buying of food on platforms, the related information sharing and crosschecking through 'authentic' food gifting and sharing, and crowdfunding activities, these are all practices that are not banal activities but instead rich in symbolic meaning. They resist the apparatus that reproduces the unsafe and 'fake' food in China, they help customers in a search for a better life. To just pigeonhole these practices as consumerism and instances of platform capitalism is a gross simplification, instead, they are more like acts of resistance, mobilising food and cuisine – part of popular culture[9] or engaging it with a sense

8 Joseph R. Gusfield and Jerzy Michalowicz, 'Secular Symbolism: Studies of Ritual, Ceremony, and the Symbolic Order in Modern Life', *Annual Review of Sociology* 10.1 (1984): 417-35.

9 John Fiske, *Reading the Popular* (London: Unwin Hyman, 1989).

of 'weapon of the weak'[10]. These are the micro tactics of everyday life,[11] creatively articulating a network of an anti-discipline[12] or an anti-dominance.

There is creativity involved in both selling (comprising ways of informing and engaging the buyers), and buying (consisting of the ways customers use to reach and engage the sellers). The digitisation and platformisation of 'natural' and 'local' operates on two creative levels: at the micro level, it enables to creatively take care of the self; at the macro level, it is a creative practice to create and maintain an apparatus of anti-dominance.

Note: This project is partly funded by China's Ministry of Human Resources and Social Security's Funding Initiative for Returned Overseas Students and Scholars.

Bibliography

de Certeau, Michel. *The practice of everyday life* (Berkeley, CA; Los Angeles: University of California Press, 2002 [1984]).

Fiske, John. *Reading the Popular* (London: Unwin Hyman, 1989).

Gusfield, Joseph R., and Jerzy Michalowicz. 'Secular Symbolism: Studies of Ritual, Ceremony, and the Symbolic Order in Modern Life', *Annual Review of Sociology* 10.1 (1984): 417-35.

Ministry of Commerce of the People's Republic of China (MoCPRC). 'The 2016 Report on China's Online Retail'. 2016. p. 32. Accessed August 24, 2016. http://images.mofcom.gov.cn/dzsws/201706/20170621110205702.pdf.

Poell, Thomas, and José van Dijck. 'Constructing Public Space: Global Perspectives on Social Media and Popular Contestation – Introduction', *International Journal of Communication*, 10 (2016): 226-34.

Scott, James C. *Weapons of the Weak: Everyday Forms of Peasant Resistance* (New Haven, CT: Yale University Press, 1985).

Srnicek, Nick. *Platform Capitalism* (Cambridge, UK: Polity Press, 2017).

Yang, Guobin. 'Contesting Food Safety in the Chinese Media: Between Hegemony and Counter Hegemony', *The China Quarterly* (May 2013): 1-19.

Yu, Hong. *Networking China: The Digital Transformation of the Chinese Economy* (Champaign, IL: University of Illinois Press, 2017).

10 James C. Scott, *Weapons of the Weak: Everyday Forms of Peasant Resistance* (New Haven, CT: Yale University Press, 1985).

11 Michel de Certeau, *The Practice of Everyday Life* (Berkeley, CA; Los Angeles: University of California Press, 2002 [1984]).

12 Ibid., xv.

About the Author

Zeng Guohua is Assistant Professor at the Institute of Journalism and Communication, Chinese Academy of Social Sciences, Beijing, China. He received his PhD from the University of Amsterdam with a study on the Beijing Olympics, in particular the global mediation of its Opening Ceremony. He has published on Chinese internet, Chinese popular culture, as well as on rural development, and is currently working on a research project on the social and political implications of new media and materiality in China.

3.4 Participatory Art On-off a Digital Platform

A Mobius Strip: On *Cyber Nails* in *Curated Nails*

Zoénie Liwen Deng

Abstract
How does participatory art in China make use of digital platforms? What kinds of relationships are triggered and enabled by the digital and digitalisation of participatory art projects? By analysing *Cyber Nails* (2014), a subproject of the participatory art project *Curated Nails* (2014-2016), this field note explores the chiasm, the Mobius strip-like relationship between online and offline, between showing and seeing. It shows how participants in the project enter into the uncanny mirroring relationship between the corporeal self and the digital self/other, and how they are involved in a cycle of a desire to see and to be seen.

Keywords: participatory art, digital, corporeal, chiasm, platform, China

If you meet a Chinese person who has a smart phone and want to connect with them, they will probably ask: 'Will you scan me or will I scan you?' This does not literally mean that they are going to scan you, it just means that either they will scan your WeChat QR code, or vice versa. WeChat has become a popular social media application for Chinese people of different age groups: 697 million people are currently using this app.[1] WeChat is

1 前瞻网 [Qianzhanwang], '微信月活跃用户 6.97 亿 今年交易量将达 5 千亿美元 [WeChat monthly active users reached 697 million, transactions this year will reach 500 billion US dollars]',

Kloet, Jeroen de, Chow Yiu Fai, and Lena Scheen (eds), *Boredom,* Shanzhai, *and Digitisation in the Time of Creative China*. Taylor & Francis Group, 2019
DOI: 10.5117/9789462984745_CH3.4

similar to WhatsApp, but has many extra functions, e.g. people can post things in 'moments', which is more like Facebook, and organisations' official accounts can post articles regularly on their platforms, which can be shared by WeChat users. Every user can generate content on this platform, and become self-media (自媒体, *zi meiti*), a term used by Chinese columnists, critics, and intellectuals to describe the situation in which the digital and the internet enable people to use media platforms to present themselves and publicise their viewpoints, activities, images, videos and so on; everyone is able to attract attention on social media by becoming self-media.

Likewise, in contemporary China, curators and artists mobilise their projects on digital platforms, mainly using WeChat. In my PhD research project on contemporary art in China, digitisation plays an important role in promoting and calling for participation in art projects on and off digital platforms, in digital archivisation of artworks and art projects, and also in the making of artworks that are created or take place in digital spaces and only exist digitally. My research interest encompasses projects such as the socially engaged art project *5+1=6*, and participatory art project *Curated Nails*, both of which utilise the digital social media platform WeChat to publicise and mobilise their projects, and also for archivisation. *5+1=6* was launched on WeChat by the *Second Floor Publishing Institute* in Beijing, together with artist and art professor Li Yifan, in September 2014. In their open call for participation, the initiators invited cultural practitioners to 'choose one of the villages/towns between the fifth ring road and sixth ring road to conduct an investigative project in an artistic way'.[2] Some subprojects of *5+1=6* also used social media to conduct their research on inhabitants in their chosen villages, but the mediality of the digital platform was treated as an enabler and a means of producing offline results for the projects. In this field note, however, I would like to elaborate on a project that imbricates online and offline participation in various ways besides mobilisation, and manifests the interfaciality between the digital and the physical, between social media and 'meat-media' (肉媒体, *rou meiti*), a composite word used by artists in China to describe the use of the human body in media-based art works and practices.

In the same year as *5+1=6*, in September, another art project was also launched on WeChat: *Curated Nail,* by Ye Funa. Ye invited 'curators to manifest exhibition themes and concepts on the tiny space of human nails,

搜狐网 [*Sohu.com*], April 29, 2016, a cessed September 29, 2016, http://mt.sohu.com/20160429/n446764111.shtml.

2 二楼出版机构 [Second Floor Publishing Institute], '六环比五环多一环' *5+1=6 Project Launching* (Second Floor Publishing Institute WeChat platform, 2014).

breaking the barriers of 'daily display' and 'art exhibition'.[3] It started as an open call for participation on WeChat where everyone could be a curator, regardless of his or her profession. On the one hand, this project used a part of the human body as an unconventional space for an art exhibition (although it is very common in performance art for the human body to generate a performance, which renders the body as the site of gaze and affect; in cyborg art, the body itself is the site of experiment and exhibition). On the other hand, it mimicked and parodied the institutional exhibition system/white cube model on the fingernails.[4] It also had a kinship with manicure, which has been popular for the last two decades in China. Nail salons can be found in shopping malls and on the streets; articles on this year's trendy nails appear on websites; the ever-changing colourful products appear on the shelves of nail polish in shops. The project's affinity with manicure made it more accessible and amiable to people who were not in the art world, or were less 'serious' in the art world. Ye initially wanted to conduct this project in a formal way, but no art institute was willing to host the project,[5] which was one of the main reasons why she initiated the project online.

From its inauguration until November 2016, more than 100 subproject proposals were received. Some were realised offline, some were executed online, most of the time via WeChat, some took place both on- and offline, mirroring one another, and the rest remained as proposals in the digital space. The choice of execution online or/and offline for each subproject in *Curated Nails* depended on its curator. Ye herself also curated some shows and activities on and around fingernails. In this field note, I will analyse one subproject that was proposed and executed by Imo, a self-proclaimed 'net artist'. This was a 'net art' project in which the human body was nonetheless involved and situated in-between, inside and outside the digital world and the physical world. Digital art is art that uses the inherent possibilities of the digital and is produced, stored, and presented in digital format.[6]

3 叶甫纳 [Ye Funa], 指甲计划 [*Curated Nail*], powerpoint presentation at UCCA, January 15, 2015, provided by the artist.

4 'The white cube has various roots which all finally come together in the 1930s in the Museum of Modern Art in New York. Before and after the First World War, there was a desire to show pieces of art against a background with the greatest possible contrast to the dominating colours of the paintings' (Charlotte Klonk, in *The White Cube and Beyond* by Niklas Maak, Charlotte Klonk and Thomas Demand. http://www.tate.org.uk/context-comment/articles/white-cube-and-beyond, January 1, 2011).

5 Ibid.

6 Christiane Paul, *Digital Art* (*World of Art*) (London: Thames & Hudson, 2008), 3.

Internet art is a form of digital art that is made on and for the internet, also known as net art.[7] In December 2014, on WeChat, Imo launched the 'net art' project, *Cyber Nail*, in which a 'human flesh nail machine' would paint the photographed nails of a participant, using an application, once a photo of a hand (with nails) was received. One could thus have a manicure in the virtual world.[8]

Although the nails in the digital world were painted digitally, human bodies were present and involved. Curator Karen Archey analyses how the body, in some digital artworks, is mediated by both virtual and physical space, and she tries, through her analysis, to contribute to the artistic discourse by elaborating on identity, body and sexuality.[9] However, not all the cases that she chose used the human body as the space of both acting and exhibiting. In this project, if you wished to participate, you needed to first take a photo of your hand using your smart phone, during which you would use your finger to press the 'shutter' on your screen to capture the image of your other fingers. Then you would send your photo to Imo, and she would use her fingers to paint your digital nails. The interfaciality of this project is not only about the functionality of the interface of the application, but also the imbrication of the human body and the digital. Even though, as Jason Farman observes, the designers try to render invisible the computation and technologies of the digital system, to provide an 'interfaceless interface',[10] the interface between the digital world in your smart phone and your body is the screen, a physical object on which your fingers move around under your eyes.

In the case of *Cyber Nail*, when Imo painted your nails, her fingers doodled and moved on your 'fingers', your 'digital nails'; the size and shape of the brushstrokes on your 'nails' depended on the touch and movement of the fingertips of Imo. In Illustration 3.4.1, the woman in the background, with only the area around her eyes visible, is looking at those who are touching this image with their fingers and staring at it on their smart phones. In the top right corner is a series of hashtags (#hot#sexy#onlyone#lol) enticing

7 Tate, 'Internet Art', accessed December 28, 2016, http://www.tate.org.uk/learn/online-resources/glossary/i/internet-art.

8 Imo, 'Imo X 指甲计划丨 "#imo"', 展示癖 *Exhibitionist*, December 20, 2014, accessed December 10, 2016, http://mp.weixin.qq.com/s/oOUK4OM1GmNayFuToeRNZA

9 Karen Archey, 'Bodies in Space: Identity, Sexuality, and the Abstraction of the Digital and Physical', in *Mass Effect: Art and the Internet in the Twenty-first Century*, ed. Lauren Cornell and Ed Halter (Cambridge, MA; London: The MIT Press. 2015), 451.

10 Jason Farman, *Mobile Interface Theory: Embodied Space and Locative Media* (New York; London: Routledge. 2012), 7.

3.4.1 The interface between the digital world in your smart phone and your body is the screen, a physical object on which your fingers move around under your eyes

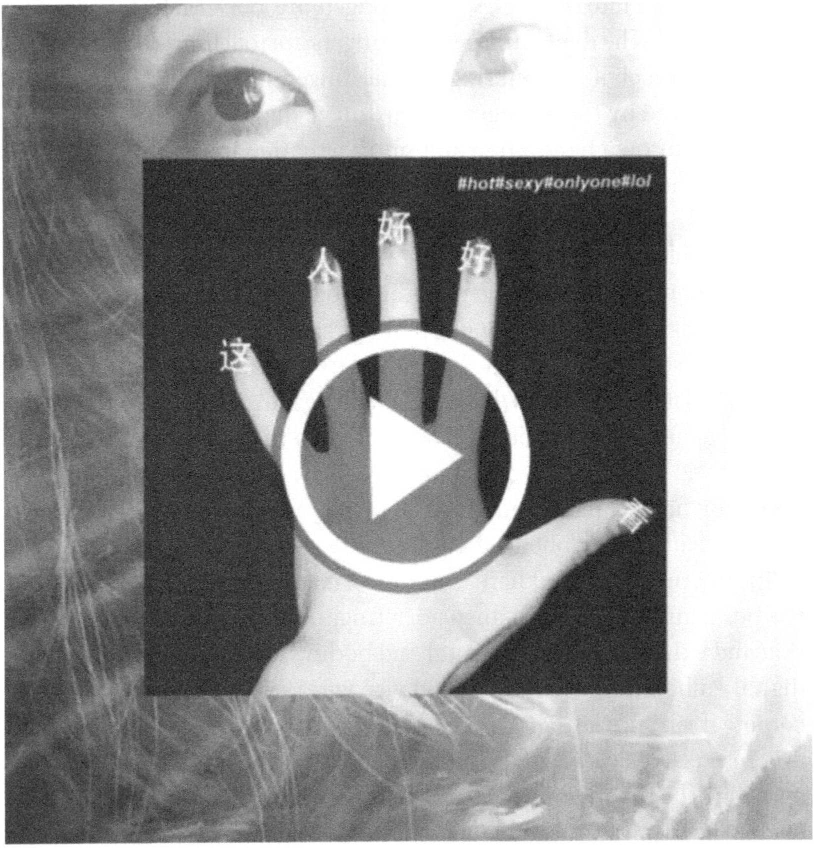

前20名分享到朋友圈有可能该视频能播放

people to touch the 'play' button with their fingers, which is at the back of a hand bearing words on the nails saying, 'this person is very good looking'. This plays with the voyeurism that is prevalent in internet culture in China: the popularity of live streaming (直播, *zhi bo*) is a good example.[11] This is also a return of the performative gaze that solicits touching and seeing. At the same time, the nails above the screen and those under the screen are 'touching' each other in an uncanny and chiasmatic way. Your fingers never touch those in the image physically, but rather, digitally; however, in order

11 See Ronald Deibert's article 'What Lies Beneath China's Live-Streaming Apps?' Available on https://deibert.citizenlab.org/2016/11/live-streaming/.

to do so, you have to use your fingers physically. 'Between my body looked at and my body looking, my body touched and my body touching, there is overlapping or encroachment, so that we may say that the things pass into us, as well as we into the things'.[12] Your fingers are looked at by both yourself and the eyes in the digital image; your fingers are touching both the screen and the digital photo of your fingers; your fingers are touched by the screen physically and by those fingers in the image digitally. If the hand in the image is your own hand, you enter into a situation in which you digitally touch your hand that has been touched and manicured by Imo digitally. The text 'The first twenty people who share this post in their WeChat moment might be able to play this video' underneath this image/video seduces you to enter a circus of collective peeping in darkness – no one knows who can watch it or who is watching it, perhaps except for Imo herself. If that is your hand, you are in/voluntarily involved in this Mobius-strip-like cycle of exhibitionist and voyeurist, looking and being looked at, and touching and being touched. All the cyber nails are stored in the cloud – on the WeChat account and website of *Curated Nails*. If you have participated in Imo's *Cyber Nail*, your 'nails' can be summoned for display with the touch of a finger on the screen of a smart phone, digitally embodying you.

To sum up, the digital here is not external or detached from the body, the flesh, and the physical. Instead, a fingernail, as a space for participation and exhibition, is both digital and bodily in an intertwined way. The digital embodies the corporeal and vice versa, blurring the binary of the corporeal self and the digital other as Moore observes.[13] Participatory digital art projects like *Cyber Nails* do not simply create an extended and expanded space for the offline online, or replicate the physical in the digital, but enable the two-way uncanny reflection and refraction of the actual and the virtual. *Cyber Nails* also echoes other Chinese contemporary art projects that use live streaming to tap into the desires for showing (exhibitionism) and watching (voyeurism) of netizens, and to explore the possibilities of exhibiting contemporary art in digital space. In response to one of Imo's nail painting (see illustration 3.4.2), I can't believe that the digital world is only approximately 60 years old!

12 Maurice Merleau-Ponty, *The Visible and the Invisible* (*Studies in Phenomenology and Existential Philosophy*) (Evanston, IL: Northwestern University Press, 1969).

13 Lorna Ann Moore, 'Be[ing] You: In[bodi]mental a Real-Time Body Swapping Video Performance', in *Handbook of Research on Digital Media and Creative Technologies*, ed. Dew Harrison (Hershey: IGI Global, 2015), 27.

3.4.2 One of the cyber nail paintings

This research is supported by a consolidator grant from the European Research Council (ERC-2013-CoG 616882-ChinaCreative).

Bibliography

Archey, Karen. 'Bodies in Space: Identity, Sexuality, and the Abstraction of the Digital and Physical'. In *Mass Effect: Art and the Internet in the Twenty-first Century,* ed. Lauren Cornell and Ed Halter (Cambridge, MA; London: The MIT Press. 2015), 451-68.

Deibert, Ronald. 'What Lies Beneath China's Live-Streaming Apps?' The Citizen Lab, Munk School of Global Affairs & Public Policy, University of Toronto. November 1, 2016. Accessed December 10, 2016. https://deibert.citizenlab.org/2016/11/live-streaming/.

Farman, Jason. *Mobile Interface Theory: Embodied Space and Locative Media* (New York; London: Routledge. 2012).
Imo. 'Imo X 指甲计划 | " #imo", 展示癖 *Exhibitionist*, December 20, 2014. Accessed December 10, 2016. http://mp.weixin.qq.com/s/oOUK4OM1GmNayFuToeRNZA.
Merleau-Ponty, Maurice. *The Visible and the Invisible* (*Studies in Phenomenology and Existential Philosophy*) (Evanston, IL: Northwestern University Press, 1969).
Moore, Lorna Ann. 'Be[ing] You: In[bodi]mental a Real-Time Body Swapping Video Performance'. In *Handbook of Research on Digital Media and Creative Technologies*, ed. Dew Harrison (Hershey, PA: IGI Global, 2015), 18-32.
Paul, Christiane. *Digital Art* (*World of Art*) (London: Thames & Hudson, 2008).
Tate. 'Internet Art'. Accessed December 28, 2016. http://www.tate.org.uk/learn/online-resources/glossary/i/internet-art.
二楼出版机构 [Second Floor Publishing Institute]. '六环比五环多一环' *5+1=6 Project Launching* (Second Floor Publishing Institute WeChat platform, 2014).
叶甫纳 [Ye Funa]. 指甲计划 [*Curated Nail*]. PowerPoint presentation at UCCA, January 15, 2015. Provided by the artist.
前瞻网 [Qianzhanwang]. '微信月活跃用户 6.97 亿 今年交易量将达 5 千亿美元 [WeChat monthly active users reached 697 million, transactions this year will reach 500 billion US dollars]', 搜狐网 [*Sohu.com*], April 29, 2016. Accessed September 29, 2016. http://mt.sohu.com/20160429/n446764111.shtml.

About the Author

Zoénie Liwen Deng is an art writer, researcher, and translator. She is currently pursuing her PhD in Amsterdam School of Cultural Analysis, University of Amsterdam. Working as a researcher for the ERC funded project *ChinaCreative*, her PhD dissertation is provisionally entitled "The Non-Oppositional Criticalities of Socially Engaged Art Concerning Urbanisation in Contemporary China". Her research and artistic interests cover social practices, feminism, the postcolonial, art and activism, critical ways of living together. She worked as a project coordinator in Shanghai Biennale in 2012. She obtained her Master's degree in cultural studies in Goldsmiths College, University of London. She is a contributor to art media such as *Leap* and *Artforum China*.

3.5 Ongoing Digitisation and Independent Chinese Documentary

A Field Report from Beijing 2015-2016

Rowan Parry

Abstract

In this essay I briefly set forth how I experienced the impact of ongoing digitization during several field trips to Beijing in 2015 and 2016, which I made as part of my current research on independent Chinese documentary. This essay highlights three digital developments, namely digital recording, the Internet, and WeChat, which have had a significant impact on the ways in which independent documentaries are produced and distributed.

Keywords: independent cinema, film exhibition, distribution, screenings, Beijing, counterculture

Production

When tracing the developments of independent documentary filmmaking in China,[1] it is impossible to neglect the role that digital technologies have played in this field. It is accepted amongst most scholars of Chinese cinema that the emergence of conveniently sized digital cameras (DV) in the late 1990s had an immense influence on independent filmmaking practices.[2] The affordability and accessibility of DV enabled a generation of filmmakers

1 Independent Chinese documentary here refers to films that embody an 'independent spirit' and are made largely independent from Chinese commercial and state film structures. Due to limited space, I cannot go into the details of what constitutes independent documentary in China, for details on this please see Berry, Rofel, and Lu (2010) or Edwards (2015).

2 Zhang Zhen and Angela Zito, eds., *DV-Made China: Digital Subjects and Social Transformation after Independent Film* (Honolulu: University of Hawaii Press, 2015).

Kloet, Jeroen de, Chow Yiu Fai, and Lena Scheen (eds), *Boredom,* Shanzhai*, and Digitisation in the Time of Creative China*. Taylor & Francis Group, 2019

DOI: 10.5117/9789462984745_CH3.5

to create films outside of the commercial and state dominated production systems. Often using a direct cinema style, these films brought to light otherwise unseen and unheard aspects of Chinese society. The liberating effects, its specific aesthetic characteristics, and the ethical challenges brought on by DV in China has been widely discussed.[3]

In the late 2000s, ten years after DV made its introduction, Chinese DV productions had gained large scale international recognition with films like *West of the Tracks* (2003), *1428* (2009), *Last Train Home* (2009), and *Petition* (2009) winning awards at film festivals around the world. At the same time, smartphones with built in cameras made their introduction and rapidly began to spread amongst the population. At present close to 60 per cent of the Chinese population owns a smartphone, and amongst people aged eighteen to 34 this is around 85 per cent.[4] This development has given rise to a generation of people for whom owning a camera, and as such being equipped to make films (and to a much larger extent photographs),[5] is not considered to be anything out of the ordinary. Coupled with the ability to instantly disseminate, edit and manipulate these images the way in which images are produced and consumed is rapidly changing, and this also seems to be having an impact on the independent film scene in China.

An example of these developments can be found in the young filmmaker Zhao Xu (born in 1992), who made his first short film in high school and dropped out after one semester of art school in order to devote all his time to making films. Readily available filmmaking equipment enabled Zhao to experiment with filmmaking from a very young age and at only 24 years of age his film *Regarding Lambs in the City* (2015), about people and sheep living on the edges of Beijing, has been screened at multiple film festivals. Also, the ease with which films can be made influences his approach to documentary film making, and gives him a desire to do more than just capture reality. For the production of his forthcoming film *Hills and Mountains*, he is using documentary footage of his friend in order to craft a semi-fictional tale that

3 Luke Robinson, *Independent Chinese Documentary: From the Studio to the Street* (London: Palgrave Macmilan, 2013); Yimin Wang, The Amateur's Lightning Rod: DV Documentary in Postsocialist China', *Film Quarterly* 58.5 (2005): 16-26.

4 Pew Research Centre, 'Smartphone Ownership and Internet Usage Continues to Climb in Emerging Economies', February 22, 2016, http://www.pewglobal.org/files/2016/02/pew_research_center_global_technology_report_final_february_22__2016.pdf.

5 Video streaming sites contain large quantities of amateur films made with nothing more than a phone, and even though smart phones are not marketed (or often used) as professional filmmaking equipment, many models possess the capabilities to produce professional quality film recordings. Examples can be found in the Korean short *Night Fishing* (2011), filmed with an iPhone4, and the American feature film *Tangerine* (2015), which was shot entirely on an iPhone 5s.

goes beyond the main characters' experiences. The main aim of this film he says is 'to capture the dynamics between the spiritual and material world'.

Where Chris Berry and Lisa Rofel noted the contrast between how 'Chinese filmmakers and commentators valued DV's ability to capture what was happening around them in a direct and unmediated manner' and how Western theorist emphasises digital filmmaking's 'ability to manipulate what is recorded in an almost equally direct manner'.[6] The ability to manipulate and craft digital images in an age where everything is recorded, posted, mashed up, and re-posted appears to be one of the driving factors towards a fusion between these two positions and a catalyst in the rise in looser forms of documentary making and docufictions, which started in the late 2000s with titles like *Disorder* (2009), *Noise* (2007), and *Oxihide* (2005), and continues today with films like *Behemoth* (2015) and *Li Wen at East Lake* (2015).

Distribution

Besides the proliferation of digital recording devices, the most important development on the digital front has been the spread of the internet. The promises and perils of the internet in China has been a hotly debated topic for many years[7] and it is not my intention to address this debate here, but instead I would like to expand on how I personally experienced the internet in doing my research.

Technically, the internet can enable people to view, share, and discuss anything in real-time, which, judging by the number of people staring at their phones on the Beijing subway, seems to be happening continuously. However, this in no way means that independent documentaries are freely disseminated across the internet. Even though it is possible to share films online, and to find various documentaries through both video-hosting sites like YouTube, Vimeo, *Tudou*, and *Youku*, and some filmmakers, particularly those with an activist agenda, like for example Ai Weiwei and Ai Xiaoming, use all channels available to disseminate their films as widely as possible, it is not a given that everything can simply be found online.

6 Chris Berry, Lu Xinyu, and Lisa Rofel, eds., *The New Chinese Documentary Film Movement: For the Public Record.* (Hong Kong: Hong Kong University Press, 2010).

7 Guobin Yang, *The Power of the Internet in China: Citizen Activism Online* (New York: Columbia University Press, 2009); Johan Lagerkvist, *The Internet in China: Unlocking and Containing the Public Sphere* (Lund: Lund University, 2006).

Firstly, uploading a film to a Chinese based website like *Youku* might also mean that it will be removed if the content is not deemed suitable. For example, the films by LGBT activist/filmmaker Fan Popo were available on these sites but were removed. An inquiry into why the films where removed, led him to find out that no order was given to remove the films. In a subsequent court case, the court ordered that the film be allowed back online. However, at present, he said during an interview, the websites have a system in place that automatically recognises banned content (including his film) and any attempt to upload this material is immediately met with a message that it cannot be uploaded. The inability to publicly share specific types of content through Chinese video streaming services means that the potential to reach large Chinese audiences through the internet is limited for many indie filmmakers.

In recent years, several films have been released on DVD by companies like dGenerate, Fanhall Filmstudio (no longer in operation), Visible Record, Icarus, CNEX, Hubert Balls Fund, and several of these titles can still be ordered online with the prices ranging from five to 395 euros. After being released these films will usually find their way onto specialist film pirating websites and can then be illegally downloaded by people who know how to navigate these parts of the internet.

Even though I was able to find a fair amount of material online through various methods, it is only a fraction of the independent documentary titles that I know have been produced. Besides several thresholds that would prevent people from disseminating their material online, it is more common that filmmakers purposefully neither upload their films online, nor release them on DVD, and on many occasions when I have been given DVDs or private links to watch or download films this has been with the request not to further disseminate the film. This is done for a variety of reasons, but most often I have found that it is done to more tightly control the distribution of their work, either because it has been agreed with the subjects that the film will not be distributed,[8] or for the films retain some kind of value and exclusivity.

In addition to these obstacles to using the internet to find independent films, also websites which are dedicated to the discussion of independent Chinese films are hard to come by or blocked in China, for example the Chinese website *Fanhall* was shut down, the U.S.-based website of dGenerate is blocked in China, and many films like *Petition* (2009) can not be listed

8 Even films of which the filmmaker agrees not to distribute them (in China) might still find their way online though.

on the popular entertainment platform *Douban*. These examples show that even though a web-savvy documentary enthusiast will be able to find a fair amount of films and information online, in this particular case the internet is not the free and open platform that it could potentially be.

Like the documentary film is somehow rooted in material reality, it is through being present at film screenings held across the city and engaging with filmmakers and audiences that I got the best understanding of what independent documentary filmmaking in China entails. However, to find out about the screenings there is one last digital hurdle to cross: WeChat.

WeChat is a smartphone application launched in 2011; in 2016 it had over 700 million active users, and close to 93 per cent of the Beijing population uses the app.[9] The application is used for pretty much anything you could possibly do with a smart phone like chatting, sharing locations, making payments, booking taxies or restaurants, and charging the credit on your phone. The app also allows people to share 'moments' with their contacts; these could be photos, texts, news articles, or event information.

In the initial phases of my fieldwork, I did not yet fully understand what WeChat meant for promoting screenings, and it was only after attending my first screening, which was listed in an expat magazine that this began to unfold. After this screening, I joined a WeChat group dedicated to the screenings at that specific club and I began to realise that only a fraction of the screenings were listed on websites and very quickly WeChat became my main channel to receive information.

As I attended more screenings, I joined more *WeCha*t group chats related to specific screening venues, followed official accounts set up by screening groups and could see events shared by my contacts in their 'moments'. As my WeChat contact list grew and I added more subscriptions, my phone became flooded with an endless stream of information. Scrolling through all of it to pick out interesting screening events became a daily activity. Not before long, the main issue was not where to find films, but to decide which of the many screenings to visit, especially during weekends when there could be up to six events on a single day.

In essence, almost all events I visited used WeChat as the main form of promotion. Most events were shared in groups, through moments, and via subscriptions, the organisers also often urged interested people to forward this information to potentially interested people. In many ways, using WeChat as a promotion channel can be seen as a contemporary form of

9 Laurie Beaver, 'WeChat breaks 700 million monthly active users', *Business Insider*, April 20, 2016, http://www.businessinsider.com/wechat-breaks-700-million-monthly-active-users-2016-4?IR=T.

word of mouth advertising. Even though WeChat has a lot larger potential reach and is much faster than word of mouth, the excess of information being received through WeChat can make it easy to overlook stuff or simply forget it as soon as the next message pops up. Unless you are especially on the lookout for screenings it becomes just one more drop in an endless stream of messages, photos, news and events.

For most venues, using WeChat is the only way used to attract an audience for the events. Several also use the event listings on douban.com. Practically nowhere did I encounter posters or flyers advertising screenings.[10] In this sense, you might say that the entire process of promoting independent film screenings has been digitised. Many events would require people to register in advance to be sure that everybody would be able to get a seat, and simultaneously keep track of the people attending.

The downside to the heavy reliance on WeChat for promoting events is that it does not only allow the organisers to keep track of people attending but the state is also keeping track of everything that is being said and done. This did not seem to impede the use of WeChat though, of the many people I met who were involved in independent film only two do not use WeChat.

In this short essay, I have attempted to briefly highlight digital transformations in the recording and distribution of independent Chinese documentary. Digital technologies have altered independent documentary in several ways, transforming their production, aesthetics, and consumption. Besides offering new possibilities for watching and promoting films, this is not without obstacles, and what will happen with all the data being automatically collected online and through WeChat remains to be seen.

This research is supported by a consolidator grant from the European Research Council (ERC-2013-CoG 616882-ChinaCreative).

Bibliography

Beaver, Laurie. 'WeChat breaks 700 million monthly active users', *Business Insider*, April 20, 2016. http://www.businessinsider.com/wechat-breaks-700-million-monthly-active-users-2016-4?IR=T.

Berry, Chris, Lu Xinyu, and Lisa Rofel, eds. *The New Chinese Documentary Film Movement: For the Public Record.* (Hong Kong: Hong Kong University Press, 2010).

10 The only exception to this was a small cinema with daily screenings, which would print posters of the films they were showing that week and hang them in the windows of the venue.

Edwards, Dan. *Independent Chinese Documentary: Alternative Visions, Alternative Publics* (Edinburgh: Edinburgh University Press, 2015).

Fan Popo. Personal interview with author and Jeroen de Kloet. January 2016.

Lagerkvist, Johan. *The Internet in China: Unlocking and Containing the Public Sphere* (Lund: Lund University, 2006).

Pew Research Centre. 'Smartphone Ownership and Internet Usage Continues to Climb in Emerging Economies'. February 22, 2016. http://www.pewglobal.org/files/2016/02/pew_research_center_global_technology_report_final_february_22__2016.pdf.

Robinson, Luke. *Independent Chinese Documentary: From the Studio to the Street* (London: Palgrave Macmillan, 2013).

Wang, Yimin. 'The Amateur's Lightning Rod: DV Documentary in Postsocialist China', *Film Quarterly* 58.5 (2005): 16-26.

Yang, Guobin. *The Power of the Internet in China: Citizen Activism Online* (New York: Columbia University Press, 2009).

Zhao Xu. Personal interview with author. January 2016.

Zhen, Zhang, and Angela Zito, eds. *DV-Made China: Digital Subjects and Social Transformation after Independent Film* (Honolulu: University of Hawaii Press, 2015).

About the Author

Rowan Parry works as a software engineer. He is finalising his PhD at the University of Amsterdam. His doctoral research focuses on how contemporary independent Chinese cinema speaks truth to power. Rowan further holds a BA in China Studies from Leiden University and an MA in Asian Studies from the University of Oslo.

3.6 Digitisation with (in/out) Borders

Lo Yin Shan, Janet Fong, and Isaac Leung

Interviewer: Lo Yin Shan
JF: Janet Fong (HK independent curator based in Beijing)
IL: Isaac Leung (HK artist & curator/chairperson of Videotage)

Q1: How has the global culture of digitisation affected creativity of local society in general?
JF: About the effect of global culture on digitisation, I think because we now have the internet, it's much easier for people to absorb new stuff from different cultures around the world, most prominently from the West, but also that of China. The upside is that there are more possibilities, like in the past we might only be able to feed on local subject matters or ideas coming from our immediate surroundings to create, but now we can get inspiration and recourses from everywhere. But one of the downsides could be: we are in a way determined by the impact of certain trends, which are changing at a much faster rate than before; nothing seems to last.
IL: In the global context, various disruptive technologies have allowed anyone to become a retailer, be it in the hotel or taxi service industries, rendering traditional business models obsolete. Airbnb and Uber, which are the best-known examples of the 'sharing economy', have invented new forms of individual and collective participation. Apart from being trendy, what has led these companies to succeed is that they offer new ways to rethink the user's behaviour, seeing as users not only have basic needs, but personal values, ranging from worker welfare, to personal health, and even environmental issues.

Although we don't see a lot of major technological inventions in this region, the global culture informed by technologies – socially conscious,

Kloet, Jeroen de, Chow Yiu Fai, and Lena Scheen (eds), *Boredom,* Shanzhai, *and Digitisation in the Time of Creative China.* Taylor & Francis Group, 2019
DOI: 10.5117/9789462984745_CH3.6

3.6.1 Replicas of the world's most iconic landmarks in China

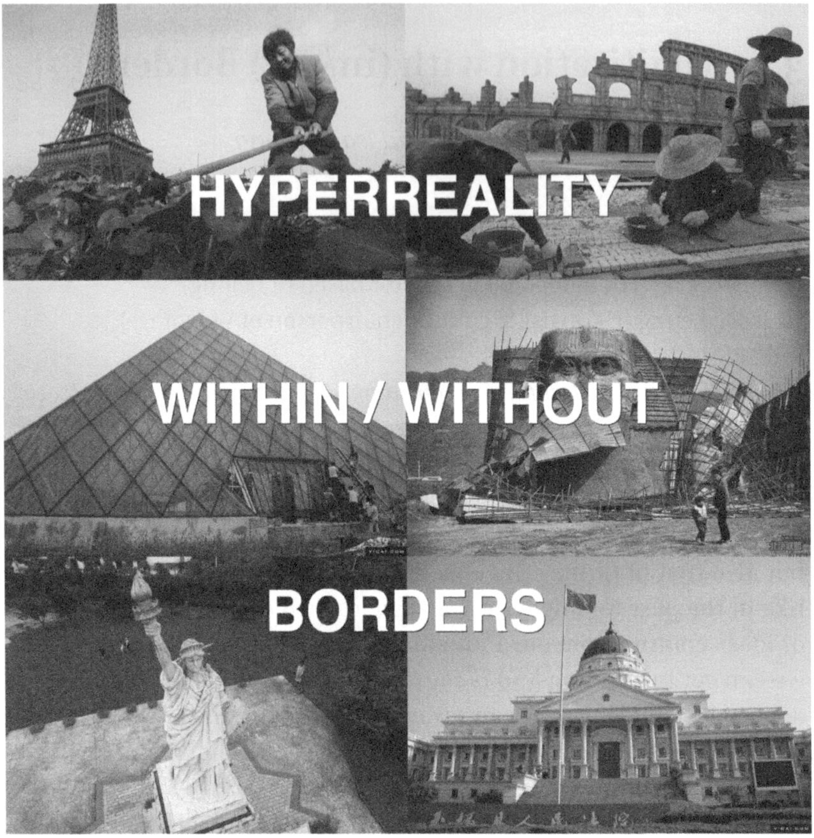

value-driven, and forward-looking – can be seen everywhere in Hong Kong and mainland Chinese societies. While some of us only use smart phones for simple tasks like taking selfies, the remaining are gamechangers in every aspect of our society. They want to experience a culture emanating from the bottom up which includes the individual's position in decision-making, voting, and contributing to any endeavour. For example, how Hong Kong millennials use social media to re-appropriate mainstream media in that it serves the political interests of the powerless is amazing.

Q2: Do you think it gives birth to new 'trends' of creativity and at the same time 'buries' the other kind of so-called tradition?
JF: The internet and the trend of digitisation surely would give birth to a lot of new things, such as new media, but at the same time, people are still relying on video as the common means of communication. In the old days,

3.6.2 The 'miniature world' from *The World* by Jia Zhangke

video might only be used for record-keeping purpose, but now video has taken over text as a more direct and effective means of distribution.

Surely, it would be great if traditions can be preserved. What traditions are at risk? Perhaps books, or written words. It's a pity that fewer people are writing by hand since the popularisation of computers and phones. But progress comes with a cost. In certain aspects, traditional handwriting has since become calligraphy, taking a new life as an art form.

For me, being a graduate of photographic studies, this stream of old school seems to be disappearing too, as all kind of morphing software like Photoshop has gradually taken the place of traditional Photography. But I would say the change is more on the technical side, the art form itself still exists. Given how easily digitised images can be distributed, it somewhat adds to the power of photography. In a way, digitisation opens up new possibilities to traditions, to truly popularise art. Of course the benchmark would inevitably be downgraded, but like I said, everything comes with a cost.

IL: I wouldn't particularly use the word 'bury'. On the contrary, creativity is essentially generative and transversal; it ties to the possibility of rediscovering the past/traditions. In this sense, I believe in Zielinski's approach of media archaeology [editorial note: in his book *Deep Time of the Media*, Siegfried Zielinski takes us on an archaeological quest into the hidden layers of media development – explores the technology of devices for hearing and seeing through two thousand years of cultural and technological history] – many near-obsolete technologies/cultures are constantly re-appropriated and re-mediated.

Q3: As China becomes a leading nation of 'Internet Plus' & 'Internet of Things' with billions of netizens, for you, what is the essence of 'Chinese style' of creativities, e.g. WeChat, *internet celebrity* (网红), *We-Media* (自媒体), *etc. which, in turn, gradually have taken control even 'outside' China? Please give some interesting example(s) about 'digitisation in China' from your personal experience.*

JF: On the Chinese style of creativities, one very impressive example is a piece of work 'Can't Help Myself', by artist couple Sun Yuan and Peng Yu from Beijing, which was exhibited in The Guggenheim Museum, New York in 2016. Inside an enclosed glass room, a giant robot (collaborated with KUKA AG) was shovelling kind of bloody liquid on the floor in 'chaotic' automation. It seemed 'uncontrollably' violent but at the same time very poetic like Chinese landscape painting.

This work was developed from their very important and timely piece, 'Freedom' (2009), a pipe fanatically watering bloody water from a fire hydrant in another enclosed space. From introspecting the relationship of freedom and their immediate society in the past, Sun Yuan and Peng Yu has transcended and moved forward to imagine our near future: the relationship of biotechnology and philosophic thinking of inner-self of human being. This piece has expressively presented the P.O.V. of Chinese contemporary art in the global context.

IL: On the one hand, the recent inventions such as instant messaging service, 'internet celebrity economy' (网红经济) and 'participatory media' could be seen as an adaptation of many existing technologies to fulfil users within the unique social and political system in China. For example, WeChat is a mixture of various popular tools in the West: part-Facebook, part-Whatsapp, and part-Amazon. On the other hand, 'micro-creativity' which corresponds to the user's daily activity is vividly manifested in the social-networking world.

For me, I don't think China has brought the world any structural changes towards (technological) innovations, so I don't think China is in any sense a leading nation in the digital realm. What interests me about the 'Chinese style of creativity' lies in how individuals survive by breaking, playing by, or re-inventing rules in their everyday life. For example, I am always amazed by how my artist and curator friends use instant messaging platforms to create their own networks and reinvent themselves. The charismatic-networked game within the social media world is highly fascinating.

Q4: In The Politics of Aesthetics: The Distribution of the Sensible, *Jacques Ranciére writes, 'It (aesthetics) is a delimitation of spaces and times, of the visible and the invisible, of speech and noise, that simultaneously determines the place and the stakes of politics as a form of experience'. How do you see the digitisation of everyday life have re-distributed the sensible and the politics of aesthetics (inside and outside China)?*

JF: 'The internet as a method' serves as a great channel of expression, especially in China – a territory of so much control and limitations. I think China is in the process of 'breaking through' its many borders, and these instances are driven by the 'united' crowd. How should I put it… this is very unlike how I have come to comprehend China, how mainland Chinese are traditionally more introverted and less likely to 'voice out', but in response to the recent smog issue, people dare to speak up. It's because when tens and thousands of people are stepping forward, people stopped being afraid. You can also see how these voices are powerful in a way that given its popular base and persistence, the PRC government can't ignore anymore and have to respond to these demands for improvement. The internet has an important role in the social structure of China, as a bridge for communication between the people and the government, which is a function no other media or platform can serve. People utilise this platform to put forward things that could not otherwise be seen or heard, I think this is a great new experience.

IL: While 'sharing economy' allows us to redistribute power and create from bottom-up, terrorists, with the same abilities and technologies, as these new business models develop new networks, could simultaneously transform themselves from a regional nuisance into a serious global threat. Although various international pro-democracy movements have been fostered through mobile technologies, we also have to admit that the vast majority of the population continues to use smart phones only for simple uses like taking selfies and 'food porn'.

'Digitisation of everyday life' could create aesthetic experiences that reframe the relation between individuals, reconfiguring how we live and how we reinvent the way in which we are prescribed for. The positive impacts of technologies can generate immense goodwill regarding individuals' sensory perception and subjectivity, but equally negative impact generated by technologies can also spiral into a full-blown crisis. That is the politics we have to face within the current landscape of the so-called digital world.

Q5: The era of big data is in certain sense about techno-democracy as well as technocracy and ubiquitous surveillance, so how do you see this paradox especially in China, e.g. internet could be viewed as an immense Intranet?
JF: In a way, this is how China differs from other countries: media in other places enjoy some degree of freedom, but in mainland China, we know there is surveillance and WeChat hires special staff to delete posts. But as this practice becomes the norm, everyone is sort of prepared about this, so when they have to say things that are likely to be censored, they would try to find other means of expression, as everyone know, this phenomenon is typically part of so-called 'Chinese characteristics'. And for example, when the government says that the air pollution is not causing any problem by getting experts to backup their claim, people would comprehend the official messages the other way round. All these phenomena of 'Chineseness' are more easily visible in the virtual space, and are exemplified by digitisation.
IL: The paradox is not uniquely Chinese; we find ourselves in a world of ubiquitous surveillance everywhere. Our personal data, usage behaviour and social network activities are constantly collected and commodified by corporations such as Facebook, Google and Amazon.
I am not optimistic about the phenomenon of mass surveillance in China, particularly because the surveillance machine has been way more powerful than the technocratic movements within the country. In recent years, technologies have been perfectly conjoined by compatible goals under the overarching political/economic agenda of the state machinery. My bigger concern is how the machine shapes the massive population's collective value in the current time.

Q6: In the past decade, even China has become the No.1 museum-booming nation, but why new media art here seems 'under-developed'? Do you think 'digitisation within border' could be one of the hindrances?
JF: Within the basic training structure of Chinese contemporary art, I think the development of new media is lagging behind. One important factor is

3.6.3 A Chinese Academy of Fine Arts as an exact imitation of 'Hogwarts School'

that in general their English competency is limited, which hinders their exposure and exploration of the latest technology – because new media, even though it has been around for a few decades already, is the newest art form, and it originates in the West. Because of the language barrier, and because new media has not been part of traditional Chinese arts education – the art schools still rely heavily on more traditional art forms to conceptualise art history, and relatively put less focus on the latest developments. My impression of new media art in China might be still video based, which is totally stuck in ten years ago. But of course they are picking up fast in recent years.

IL: Much like many contemporary arts, a lot of so-called new media art pieces in China seem to be trapped in the rhetoric of a slick artistic development. It is questionable whether anything of substance has been reflected in many new media art pieces and exhibitions. Despite the audio-visual effects of many new media art pieces being stunning, many seem to have overinflated a claim of the artistic term more to its underlying meaning. For me, the biggest hindrance is that the art world is yet to have a more philosophical/theoretical discourse about new media art. Artists and curators should avoid producing works/exhibitions that chase after technology for technology's sake.

About the Authors

Lo Yin Shan graduated with a diploma in design from the Hong Kong Polytechnic University, and is currently a feature reporter/editor and an amateur artist/photographer. She lived in Beijing from 2008 to 2017 as a 'teleporter'/columnist between two cities. Author of *Driving Lantau: A Whisper of an Island*; *Ten Notes on Northern Drifting*; editor of *The Box Book: Beauty and the Beast of Hong Kong Culture*.

Janet Fong, co-founder/ Director for IAER (International Art Exchange and Residency) in New York, is promoting awareness of art exchange and artist residency programmes by holding four panel discussions around China. She aims to focus on the development of ongoing exhibitions, events, and programmes, help build a new way for art exchange and artists to create artworks and change.

Isaac Leung is a practicing artist, curator, and scholar in art and culture. In 2003, Leung received an Honorary Fellowship of a Bachelor of Fine Arts at the New Media Art Department of the School of the Art Institute of Chicago. In 2013, Leung was appointed as the Chairman of Videotage. During his tenure, he initiated and participated in projects that included exhibitions, workshops, lectures, publications, online projects, and symposia. Some of these include 40 Years of Video Art in Germany and Hong Kong (Hong Kong and Germany), The 12th Venice Biennale International Architecture Exhibition (Italy and Hong Kong), Perpetual Art Machine (USA), Time Test – International Video Art Research Exhibition (China), ISEA Festival (Hong Kong, China), Both Sides Now (Hong Kong, UK, and various countries), Loop Barcelona (Spain), One World Exposition (Hong Kong), China Remixed (USA), Clockenflap (Hong Kong), and Art Basel Crowdfunding Initiative (Hong Kong). Leung has taught at the Chinese University of Hong Kong, the Hong Kong Baptist University, the City University of Hong Kong, the Education University of Hong Kong and the Hong Kong Polytechnic University. He is currently Assistant Professor in the Faculty of Arts of the Chinese University Biographies of contributors

Index

For Product Safety Concerns and Information please contact our
EU representative GPSR@taylorandfrancis.com Taylor & Francis
Verlag GmbH, Kaufingerstraße 24, 80331 München, Germany